# Mobile Communication Basic Concepts



*by*

A. Sindhu Devi

B.E., M.TECH., MBA., B.SC., B.ED.

Double9 BOOKS

# Mobile Communication Basic Concepts
**by A. Sindhu Devi**



ISBN: 978-93-67149-30-0

**Published by**

**DOUBLE 9 BOOKS**
2/13-B, Ansari Road
Daryaganj, New Delhi – 110002
info@double9books.com
www.double9books.com
Tel. 011-40042856

# Shape of Cells

The coverage area of cellular networks are divided into cells, each cell having its own antenna for transmitting the signals. Each cell has its own frequencies. Data communication in cellular networks is served by its base station transmitter, receiver and its control unit.

The shape of cells can be either square or hexagon –

**Square**

A square cell has four neighbors at distance d and four at distance Root 2d

- Better if all adjacent antennas equidistant
- Simplifies choosing and switching to new antenna

**Hexagon**

A hexagon cell shape is highly recommended for its easy coverage and calculations. It offers the following advantages –

- Provides equidistant antennas
- Distance from center to vertex equals length of side

## Frequency Reuse

Frequency reusing is the concept of using the same radio frequencies within a given area, that are separated by considerable distance, with minimal interference, to establish communication.

Frequency reuse offers the following benefits –

- Allows communications within cell on a given frequency
- Limits escaping power to adjacent cells
- Allows re-use of frequencies in nearby cells
- Uses same frequency for multiple conversations
- 10 to 50 frequencies per cell

For example, when N cells are using the same number of frequencies and K be the total number of frequencies used in systems. Then each cell frequency is calculated by using the formulae K/N.

In Advanced Mobile Phone Services (AMPS) when K = 395 and N = 7, then frequencies per cell on an average will be 395/7 = 56. Here, cell frequency is 56.

# Cellular Concepts

Cellular network is an underlying technology for mobile phones, personal communication systems, wireless networking etc. The technology is developed for mobile radio telephone to replace high power transmitter/ receiver systems. Cellular networks use lower power, shorter range and more transmitters for data transmission.

## Features of Cellular Systems

Wireless Cellular Systems solves the problem of spectral congestion and increases user capacity. The features of cellular systems are as follows –

- Offer very high capacity in a limited spectrum.
- Reuse of radio channel in different cells.
- Enable a fixed number of channels to serve an arbitrarily large number of users by reusing the channel throughout the coverage region.
- Communication is always between mobile and base station (not directly between mobiles).
- Each cellular base station is allocated a group of radio channels within a small geographic area called a cell.
- Neighboring cells are assigned different channel groups.
- By limiting the coverage area to within the boundary of the cell, the channel groups may be reused to cover different cells.
- Keep interference levels within tolerable limits.
- Frequency reuse or frequency planning.
- Organization of Wireless Cellular Network.

Cellular network is organized into multiple low power transmitters each 100w or less.

# CONTENTS

# ABOUT THE AUTHOR

The author of this book is Mrs. A.Sindhu Devi. She is working as an Assistant Professor in NEW PRINCE SHRI BHAVANI COLLEGE OF ENGINEERING AND TECHNOLOGY,GOWRIVAKKAM,CHENNAI. She has done her schooling in various parts of Tamilnadu. She has completed her undergraduate degree in Electronics and Communication with First class in an Anna University affiliated college. She has completed her MBA degree under Alagappa University with First class After her marriage she took a break in her studies and completed B.Sc in Mathematics B.Ed in Mathematics. She joined M.Tech in Computer science Engineering in BIHER, Chennai. She had worked in matriculation and CBSE schools as Science, Computer Science and Mathematics teacher. She had published papers on Cloud Computing and Network security on journals. Due to her interest in Computer science and her kids and husband's support she has been working on to write a book on Mobile Computing. Her husband and kids has been her backbone for her academic journey and work journey. The major reason for her author journey is her husband Mr. Venkadesh and her two kids Selvi. Vaishnavi Devi and Selvi. Saindhavi Devi.

# Cell Splitting And Cell Sectoring

With the growing number of mobile users, it is important for the cellular capacity to also keep growing to meet the needs of the users. Figuratively speaking there are broadly two ways to increase the channel capacity:

1. The new addition of channels
2. Borrowing of frequency

Both of the above two methods will also lead to an increase in cost along with capacity. Two distinct approaches we use in the modern day to increase channel capacity are Cell Splitting and Cell Sectoring. We will discuss them in detail below.

## Cell Splitting

Cell Splitting is the process of subdividing a cell into smaller cells each with its own Base Station. On splitting, new cells with smaller radius are added called microcells. Each new cell created is independent and has reduced antenna height and transmitter power. The creation of new smaller cells increases the capacity of the system as a whole. Cell Splitting increases the frequency reuse factor. A higher frequency reuse factor increases the capacity of the cellular system in Cell Splitting.

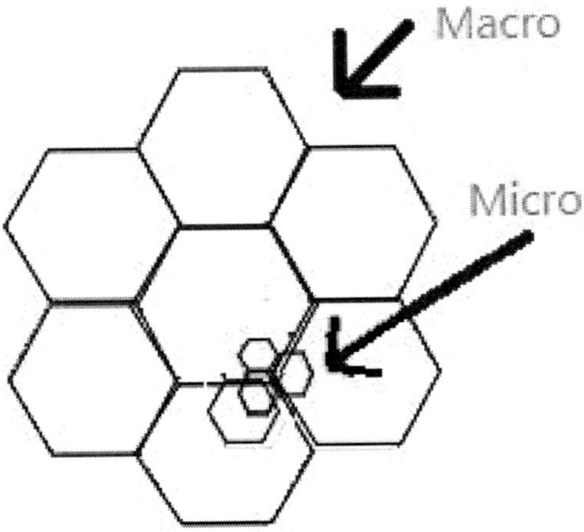

**Advantages**

- Increases the capacity of the channel considerably.
- Enhances dependability of cellular networks.
- Increases the frequency reuse factor.
- Increases signal-to-noise (SNR) ratio.
- Reduces interference.

**Disadvantages**

- For each individual cell, an individual base station is required so a huge number of base stations are needed in this process.
- Handoff occurs frequently.
- Assigning channels is difficult

## Cell Sectoring

Cells are divided into a number of wedge-shaped sectors, each with its own set of channels. By wedge-shaped we mean that the cells are divided at an angle of 120° or 60°. These sectored cells are called microcells. Like Cell Splitting, it also helps in increasing channel capacity and decreases channel interference. 3 or 6 sectors are created from a given cell. But unlike Cell Splitting, here the cell radius does not change after sectoring the cells although the co-channel reuse ratio has decreased. It increases system performance by using a directional antenna.

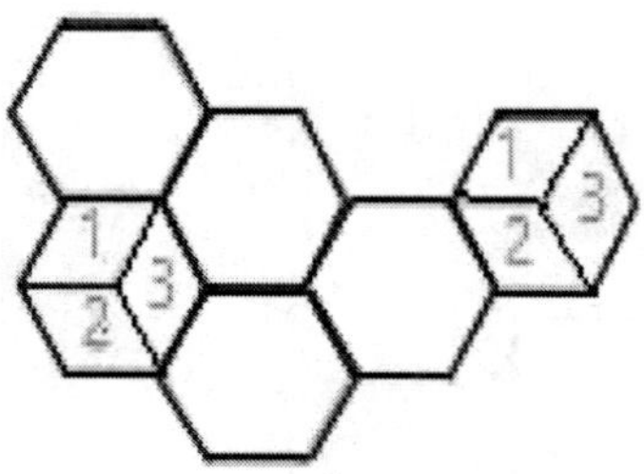


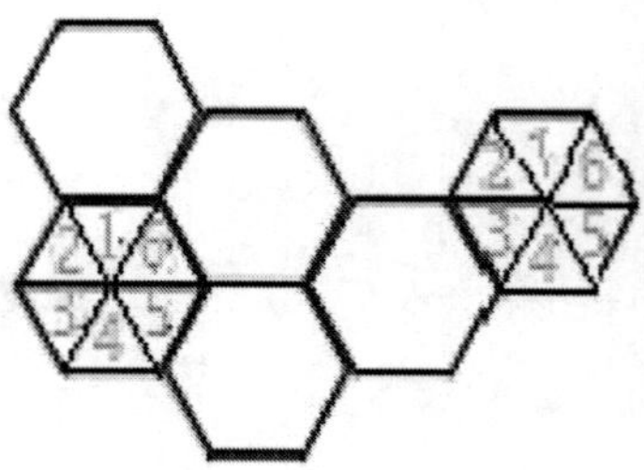


**Advantages**

- Sectoring increases the signal-to-interference ratio which means the cluster size gets reduced.
- Reduces interference without altering the system performance.

- Increases channel capacity without necessarily changing the cell radius.
- Increases frequency reuse by reducing the number of cells in the cluster.
- Assigning a channel is easier.

### Disadvantages

- Increases the number of antennas per base station.
- It decreases efficiency as sectoring reduces the channel groups.
- Excessive interference leads to traffic loss.
- The number of handoffs increases as the working area of the cell decreases in Cell Sectoring.

# Comparison between Cell Splitting and Cell Sectoring

The following table highlights the major differences between Cell Splitting and Cell Sectoring –

| Basis of comparison | Cell Splitting | Cell Sectoring |
|---|---|---|
| Definition | A method in which an area of a cell or an independent coverage area of a cellular system is divided into two or more cell areas in order to accommodate more than one cell. | A method in which a single cell is partitioned into a number of wedgeshaped sectors, each of which contains its own unique set of channels. |
| Function | This will result in a higher processing burden as well as an increased number of handoffs (base station transfers) each call. | This feature reduces the efficiency of the trunking process by dividing the channel sets into smaller groupings. |
| Operation | The cell radius is shrunk but the co-channel reuse ratio is maintained at the previous level. | The cell radius does not change, despite the fact that the co-channel reuse ratio has reduced. |

# Channel Allocation Strategies

Channel Allocation means to allocate the available channels to the cells in a cellular system. When a user wants to make a call request then by using channel allocation strategies their requests are fulfilled. Channel Allocation Strategies are designed in such a way that there is efficient use of frequencies, time slots and bandwidth.

## Types of Channel Allocation Strategies

These are Fixed, Dynamic, Hybrid Channel Allocation and Borrowing Channel Allocation as explained as following below.

### Fixed Channel Allocation (FCA):

Fixed Channel Allocation is a strategy in which fixed number of channels or voice channels are allocated to the cells. Once the channels are allocated to the specific cells then they cannot be changed. In FCA channels are allocated in a manner that maximize Frequency reuse.

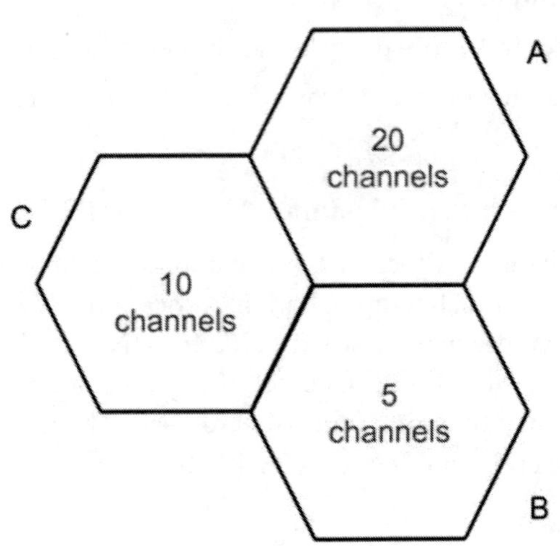


In cell A 20 Channels or Voice channels are allocated. If all channels are occupied and user make a call then the call is blocked. *Borrowing Channels* handles this type of problem. This cell borrow channels from other cells.

**Advantages :**

1. Simple to implement and manage
2. Does not require complex equipment or algorithms

**Disadvantages :**

1. Limited channel utilization as unused channels remain unused.
2. Susceptible to interference and congestion.

## Dynamic Channel Allocation (DCA)

Dynamic Channel allocation is a strategy in which channels are not permanently allocated to the cells. When a User makes a call request then Base Station (BS) send that request to the Mobile Station Center (MSC) for the allocation of channels or voice channels. This way the likelihood of blocking calls is reduced. As traffic increases more channels are assigned and vice-versa.

**Advantages :**

1. Efficient use of available bandwidth.
2. Reduces call blocking and improves call quality.
3. Allows for dynamic allocation of resources.

**Disadvantages :**

1. Requires more complex equipment and algorithms.
2. May result in call drops or poor quality if resources are not available

## Hybrid Channel Allocation (HCA)

Hybrid Channel Allocation is a combination of both Fixed Channel Allocation (FCA) and Dynamic Channel Allocation (DCA). The total number of channels or voice channels are divided into fixed and dynamic set. When a user make a call then first fixed set of channels are utilized but if all the fixed sets are busy then dynamic sets are used. The main purpose of HCA is to work efficiently under heavy traffic and to maintain a minimum S/I.

**Advantages :**

1. Provides the benefits of both FCA and DCA.
2. Allows for dynamic allocation of resources while maintaining predictable call quality and reliability.

**Disadvantages :**

1. Requires more complex equipment and algorithms than FCA.
2. May not provide the same level of efficiency as pure DCA.

## Borrowing Channel Allocation (BCA)

When a cell experiences high traffic demand and all of its channels are occupied, it can borrow channels from neighboring cells that are not being used at that time. The borrowed channels are assigned to the busy cell and are used to support the additional traffic demand. Once the demand subsides, the borrowed channels are released and returned to their home cell. BCA can be implemented manually or automatically using algorithms or policies but the main disadvantage is that if the borrowed channel is reclaimed by the original cell the call drop may occur.

**Advantages :**

1. Efficient use of available bandwidth.
2. Reduces call blocking and improves call quality.

**Disadvantages :**

1. Increases interference between cells.
2. Can cause call drops if borrowed channels are reclaimed by the home cell.

# Handoff in Cellular Telecommunications

In cellular telecommunications, the terms handover or handoff refers to the process of transferring an ongoing call or data connectivity from one Base Station to another Base Station. When a mobile moves into a different cell while the conversation is in progress then the MSC (Mobile Switching Center) transfers the call to a new channel belonging to the new Base Station.

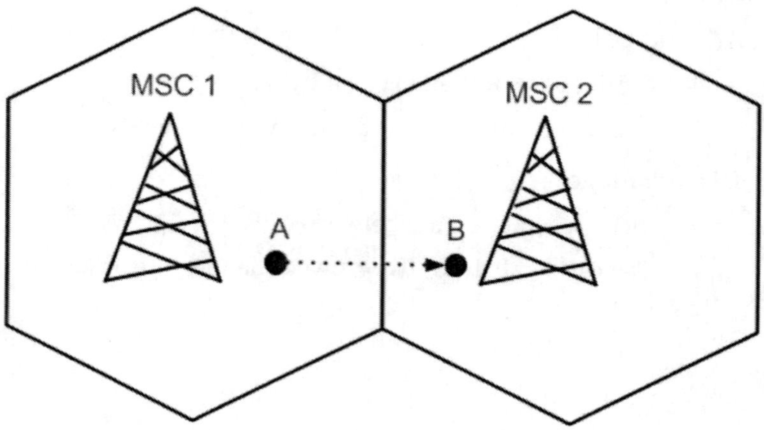


When a mobile user A moves from one cell to another cell then BSC 1 signal strength loses for the mobile User A and the signal strength of BSC 2 increases and thus ongoing calls or data connectivity for mobile users goes on without interrupting.

### Types of Handoff

- Hard Handoff
- Soft Handoff
- Delayed Handoff
- Mobile-Assisted Handoff

### Hard Handoff

When there is an actual break in the connectivity while switching from one Base Station to another Base Station. There is no burden on the Base Station and MSC because the switching takes place so quickly that it can hardly be noticed by the users. The connection quality is not that good. Hard Handoff adopted the 'break before make' policy.

It is generally implemented in Time Division Multiplexing and Frequency Division Multiplexing when a user connects to the base station with a fluctuating radio frequency.

- Hard Handoff is cheaper in cost as compared to soft Handoff because only one channel needs to be active at a time.
- It is more efficient than soft handoff, that's why hard handoffs are widely implemented.
- Sometimes, a delay can be experienced while switching base stations.

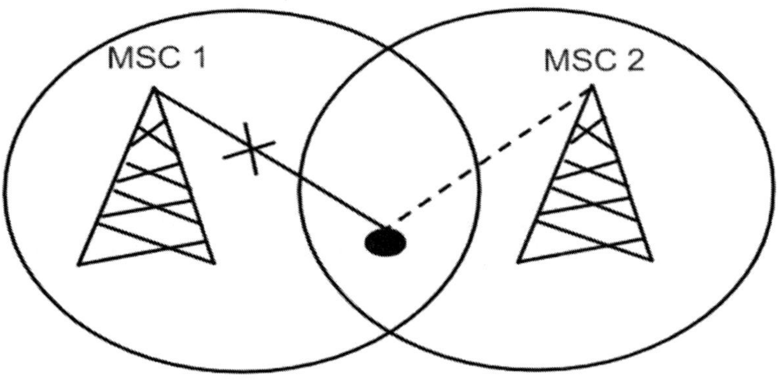


### Soft Handoff

Soft Handoff is a mechanism in which the device gets connected with two or more base stations at the same time. At least one of the links is kept when radio signals are added or removed to the Base Station. Soft Handoff adopted the 'make before break' policy. If a channel is in power loss then another channel will always be on standby mode so this makes it best in terms of quality as compared to Hard handoff. Soft handoffs are used in devices supporting CDMA/WDMA networks

- High Transmission speed as more than one repeater can transmit signals.

- It has a very low delay in signals.
- It can't be implemented on devices supporting GSM or LTE networks.

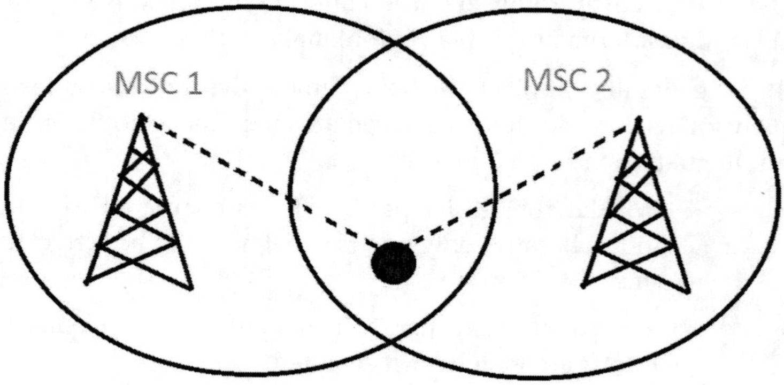


### Delayed Handoff

Delayed handoff occurs when no base station is available for accepting the transfer. The call continues until the signal strength reaches a threshold, and after that, the call is dropped. Generally, it happens when the user is out of the network coverage area, or at some dead spots where network reach is very low.

### Mobile-Assisted Handoff

Mobile-Assisted handoff is generally used when a mobile phone helps a base station to transfer the call to another base station with better-improvised connectivity and more signal strength. This handoff is used in TDMA technique-based GSM devices.

# Interference

Interference is the major limiting factor in the performance of cellular radio systems.

Sources of Interference are:

- Another mobile in the same cell.
- A call in progress in the neighbouring cell.
- Other Base Station operating in the same frequency band
- Any other non cellular system which leaks energy into cellular frequency band. Effect of interference on voice channels results in cross talks. Effect on control channels results in dropped calls or missed calls. The two major types of system-generated cellular interference are *co-channel interference and adjacent channel interference.*

## Co-Channel and Adjacent Channel Interference

Co-channel interference is not actually an interference but more a sort of congestion when more than one device is operating on the same frequency channel. It hinders performance by increasing the wait time as the same channel is used by different devices. Sources of interference are as follows:

- Another mobile in the same cell.
- Ongoing call in neighbor's cell
- When a different Base Station operates on the same frequency.

Types of channels on which interference is present are-

1. On Voice Channel: leads to crosstalk which is an interference or background noise while we are talking to another person on mobile. Crosstalk is unwanted interference that should be minimized.

2. On Control Channel: It can lead to problems in creating a connection between the sender and receiver during a call which leads to missed calls. Calls may terminate abruptly known as blocked calls.

Interference causes the above two problems because it reduces the channel capacity and thus, affects the performance.

Types of Interference in Mobile Communication

1. Co-Channel Interference
2. Adjacent Cell Interference

## Co-Channel Interference

Co-channel cells are those cells that use the same frequency in a given coverage area. Interference from these cells is called co-channel interference In co-channel interference, the cells are clustered as close together as possible to reduce the co-channel interface and provide sufficient isolation. Increasing the co-channel reuse ratio improves the transmission quality because of the smaller level of co-channel interference. An example of co-channel interference is when a radio transmitter is operating on the same frequency.

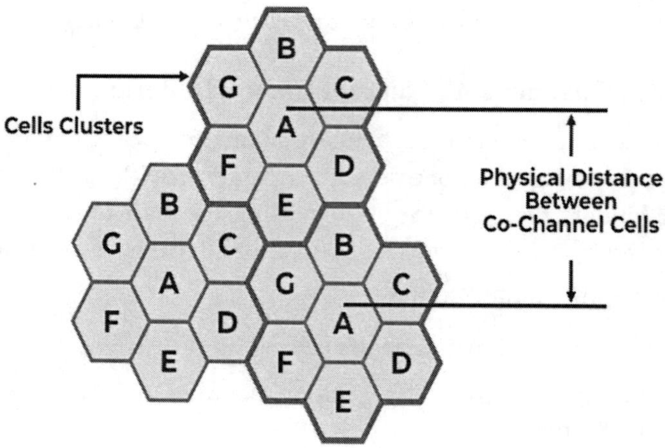


The reasons behind Co-channel interference are:

- Bad weather condition
- Poor frequency planning

Ways we can reduce co-channel interference in cellular communication are:

- Proper planning and implementation.
- The frequency reuse technique increases overall system capacity.

## Adjacent Channel Interference

It is the interference caused to the signal which is adjacent in frequency to the desired signal. Imperfect receiver side filters allow the neighboring signal to mix with the actual pass band. if adjacent channel signal strength becomes strong, it will be difficult for Base Station to differentiate the actual mobile signal from the strong mobile signal.

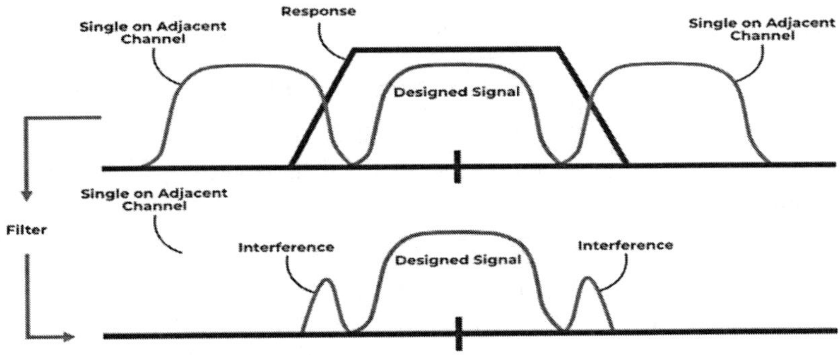


The reasons behind adjacent channel interference are as follows:

- Due to multiple channels close to each other communicating using similar frequencies.
- Irrelevant power emission from an adjacent channel.

Factors for reducing Adjacent Channel Interference are as follows:

- Proper filtering
- Careful Channel Assignments
- By managing the space between two adjacent cells which should remain constant.

# Capacity of a channel

By capacity of a channel, it means the capacity of the transmission medium (wire or link). Capacity is the number of bits the transmission medium can hold. So basically there are 2 types of channels – Full duplex and half duplex.

1. Half duplex – the transmission can happen in one direction at a time.
2. Full duplex – the transmission can happen in both the direction simultaneously.

For example, the transmission medium is operating in its maximum capacity then at that time the number of bits it is holding is called capacity of the transmission medium.

- If the length of the transmission medium is longer than its capacity will be higher.
- It also depends on the area of cross section of the medium.
- If the bandwidth is 1 bps, then every second it can take 1 bit. After every second it will move forward so that next bit could occupy the space. Therefore the final time in which it will occupy all the bits will be its propagation delay.

The capacity of the channel depends on two things:

1. Bandwidth
2. Propagation delay

Capacity = bandwidth * propagation delay

(in case of half duplex)

Capacity =2* bandwidth * propagation delay

(in case of full duplex)

## Overview of available cellular standards

The G in 2G, 3G, 4G or 5G stands for "Generation". The very first generation of a commercial cellular network, 1G, was developed since the late 1970s and introduced in 1987 by Telstra, an Australian operator. On a frequency of 30 kHz and a bandwidth of 2 kilobit per second (kbps) it was a technology that came with very poor battery life, low voice quality and not much security. It did not survive long.

## 2G - The first digital standard

2G has been established in 1991 and is the first digital standard. It delivered more reliable and secure communication. The 2G standard implemented the CDMA and GSM concepts:

- CDMA (Codedivision multiple access) enables several transmitters to send information simultaneously. Basically, it allows several users to share a band of frequencies.
- GSM (Global System for Mobile Communications) is a standard to describe the protocols for second-generation digital cellular networks. Developed in the early 1990s in Europe it became a global standard by the mid-2010s.

2G networks introduced many of the fundamental mobile services we still use today. One of them: The Short-Message-Service SMS. SMS was doing machine-to-machine (M2M) communication long before the term “internet of things” made its rounds. With “General Packet Radio Service” (GPRS) and “Enhanced Data Rates for GSM Evolution” (EDGE) the 2G Standard was updated twice to 2.5G and 2.75G enhancing its maximum speed to 171 kbps and 384 kbps.

2G connectivity is up to now a reliable standard for connecting low-bandwidth devices with the Internet of Things. However, the days of 2G are counted. Since newer standards have emerged, Telstra was one of the first operators abandoning 2G in December 2016. In Europe the 2G phase out will start by the end of 2020.

## 3G - Introducing mobile multimedia capabilities

3G was introduced in 2003. At its core network it used a new architecture called “Universal Mobile Telecommunications System” (UMTS). Its main advancement to its predecessor 2G was a significantly higher bandwidth. This made 3G the first mobile “multi-media standard”. For the first time, video streams could be transmitted via mobile cellular networks. In 2006 3G was updated to 3.5G by introducing the High-Speed Downlink Packet Access (HSDPA) communication protocol further increasing the data bandwidth of 3G up to 42 Mbps.

Another update introduced the "Long Term Evolution" (LTE) standard that increased the capacity and speed of mobile communications once more. Even though LTE has been marketed as "4G" or "4G LTE" it does not meet the technical criteria of "real" 4G wireless service. Technically LTE is a "3.9G".

However, the days of 3G are counted as well. The phase out has already begun. By mid-2021, Vodafone plans to shut down its last 3G stations in Germany. Telefónica and Deutsche Telekom have also started to remove 3G stations in favour of newer technologies.

### 4G - High speed and ubiquitous computing

The fourth Generation 4G was introduced in 2012. Its main purpose is to deliver high speed communication with enhanced security to enable high definition mobile TV, video conferencing and pervasive computing respectively ubiquitous computing with bandwidths up to 150 Mbps. While 4G sets the standards and general conditions, LTE delivers the technological fundament to fulfil these standards. But only the enhanced LTE+ or LTE Advanced is meeting the 4G requirements.

### NB-IoT and LTE-M

In connection with 4G, two more technologies emerged: NB-IoT and LTE-M. Both specifically designed for machine to machine communication. The NB in NB-IoT stands for Narrow Band and utilizes radio waves, which enable particularly large area coverage. At the same time, they can penetrate thick concrete walls and thus reach even remote corners of a building. NB-IoT is perfectly designed for constrained devices that usually transmit small data packets only once an hour or a day , making them perfect to run on batteries over a very long period. LTE-M however delivers higher bandwidths with lower latencies than NB-IoT but comes with a slightly higher energy consumption.

## 5G - The future standard

5G is the latest standard in cellular communication. Its worldwide deployment has just begun. It is estimated that by 2023 up to 32 percent of North America mobile connections will work on a 5G network. Also, in Europe 5G rollout is heavily underway. With bandwidths of up to 1 Gbps 5G is designed to enable high speed communication with high capacities and very low latency making it the perfect communication technology for augmented reality applications, gaming, machine to machine communication and smart devices when quick reaction plays a vital role. For the time being 5G is not designed to replace its predecessor 4G. The parallel operation of both technologies will enable larger capacities and faster network speeds to be served in the future.

# WIRELESS STANDARDS

- Wi-Fi standards are constantly changing and evolving, with updates bringing faster internet speeds, better connections, and more simultaneous connections.
- The most commonly used Wi-Fi standard at present is 802.11ac, while the next generation standard, Wi-Fi 6 (802.11ax), is being rolled out.
- Wi-Fi 6E is an extension of Wi-Fi 6 that allows for broadcasting over a new 6GHz band, increasing available network capacity and reducing interference. Wi-Fi 7 and Wi-Fi 8 are also on the horizon, with even faster speeds and improved features.

Wi-Fi is a catch-all term. In a sense, it is very precise. It explains a specific method you can use to connect to the internet.

There are many different types of Wi-Fi standards. Your router, laptop, tablet, smartphone, and smart home devices use different wireless standards to connect to the internet. Wireless standards change every few years, too. Updates bring faster internet, better connections, more simultaneous connections, and so on—but the sheer number of wireless standards and specifications is confusing, to say the least.

## What Are Wi-Fi Standards?

Wireless standards are a set of services and protocols that dictate how your Wi-Fi network (and other data transmission networks) acts.

The most common wireless standards you will encounter are the IEEE 802.11 Wireless LAN (WLAN) & Mesh. The IEEE updates the 802.11 Wi-Fi standard every few years. At the time of writing, the most commonly used Wi-Fi standard is 802.11ac, while the next generation Wi-Fi standard, 802.11ax (also known as Wi-Fi 6 and Wi-Fi 6E)

Now, the generation after 802.11ax is on the horizon, with IEEE 802.11be mooted for launch around 2024/2025 (using the name Wi-Fi 7). Before you ask, yes, Wi-Fi 8 is under development, currently using the name IEEE 802.11bn, but it's not expected to launch until at least 2028, and it'll likely be a few years after that, given how Wi-Fi launches and uptake tend to filter out slowly.

**A Brief History of Wireless Standards**

| IEEE Standard | Wi-Fi Alliance Name | Year Released | Frequency | Maximum Data Rate |
|---|---|---|---|---|
| 802.11a | Wi-Fi 1 | 1999 | 5GHz | 54Mbps |
| 802.11b | Wi-Fi 2 | 1999 | 2.4GHz | 11Mbps |
| 802.11g | Wi-Fi 3 | 2003 | 2.4GHz | 54Mbps |
| 802.11n | Wi-Fi 4 | 2009 | 2.4GHz & 5GHz | 600Mbps |
| 802.11ac | Wi-Fi 5 | 2014 | 2.4GHz & 5GHz | 1.3Gbps |
| 802.11ax | Wi-Fi 6 | 2019 | 2.4GHz & 5GHz | 10-12Gbps |
| 802.11ax-2021 | Wi-Fi 6E | 2021 | 2.4GHz, 5GHz, & 6GHz | 10-12Gbps |
| 801.11be | Wi-Fi 7 | 2024/2025 | 2.4GHz, 5GHz, & 6GHz | 40Gbps |

Not all old Wi-Fi standards are obsolete. At least, not yet. Here is a brief history of Wi-Fi standards and whether the standard is still active.

- IEEE 802.11: The original! This now-defunct standard was created in 1997 and supported a blazing-fast maximum connection speed of 54 megabits per second (Mbps). Devices using this haven't been made for over a decade and won't work with today's equipment.
- IEEE 802.11a: Created in 1999, this version of Wi-Fi works on the 5GHz band. This was done with the hope of encountering less interference, since many devices (like most wireless phones) also use the 2.4GHz band. 802.11a is fairly quick, too, with maximum data rates topping out at 54Mbps. However, the 5GHz frequency has more difficulty with objects in the signal's path, so the range is often poor.
- IEEE 802.11b: Also created in 1999, this standard uses the more typical 2.4GHz band and can achieve a maximum speed of 11Mbps. 802.11b was the standard that kick-started Wi-Fi's popularity.
- IEEE 802.11g: Designed in 2003, the 802.11g standard upped the maximum data rate to 54Mbps while

retaining the reliable 2.4GHz band. This resulted in the widespread adoption of the standard.

- IEEE 802.11n: Introduced in 2009, this version had slow initial adoption. 802.11n operates on both 2.4GHz and 5GHz, as well as supporting multi-channel usage. Each channel offers a maximum data rate of 150Mbps, which means the standard's maximum data rate is 600Mbps.
- IEEE 802.11ac: The ac standard is what you will find most wireless devices using at the time of writing. Initially released in 2014, ac drastically increases the data throughput for Wi-Fi devices up to a maximum of 1,300 megabits per second. Furthermore, ac adds MU-MIMO support, additional Wi-Fi broadcast channels for the 5GHz band, and support for more antennas on a single router.
- IEEE 802.11ax: Next up for your router and your wireless devices is the ax standard. As 802.11ax completes its rollout, you will have access to a theoretical network throughput of 10Gbps—around a 30-40 percent improvement over the ac standard. Furthermore, wireless ax will increase network capacity by adding broadcast subchannels, upgrading MU-MIMO, and allowing more simultaneous data streams.
- IEEE 802.11be: Although the specifications for 802.11be are yet to be finalized, it is highly likely that it will become the successor to 802.11ax. As per the IEEE Xplore paper, 802.11be will deliver "doubled bandwidth and the increased number of spatial streams, which together provide data rates as high as 40 Gbps."

# Signal

The physical representation of data by the transmission of which communication takes place is known as signal. The representation can be electrical, electronic or optical. Voice, images, letters or numbers can be represented using signals. Signal can be analog or digital.

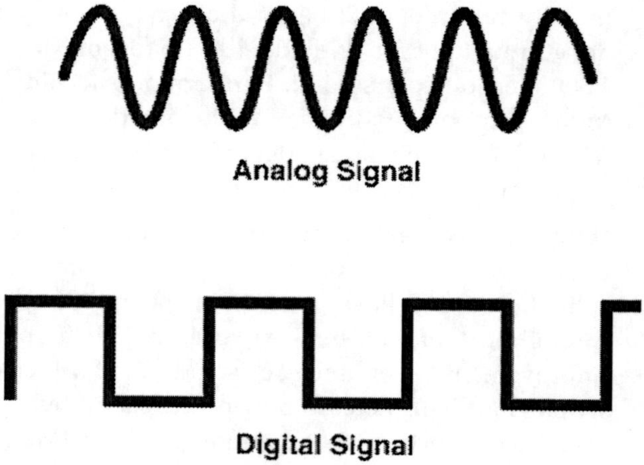


Figure- Analog Signal v/s Digital signal

Factors affecting strength and path of EM waves

- Distance between transmitter and receiver
- Frequency/wavelength of wave
- Earth's atmosphere
- Environmental objects

**Signal Propagation modes**

A signal on its journey from transmitter to receiver propagates in three modes. Type of Propagation is determined by the carrier frequency.

- Ground waves or surface waves propagation
- Sky waves or Ionospheric propagation
- Space waves or Line-of-sight propagation

**Ground Waves Propagation also known as Surface Wave (<2MHz)**

Waves with low frequencies follow the earth's surface. Due to very less frequency range, also known as low frequency/medium waves. Antennas are bigger in size and located near the ground. They follow curvature of the Earth. These waves are used in Submarine Communication, A.M radio

Figure- Ground Waves Propagation

Why the Electromagnetic waves in this frequency band follow the earth's curvature?

- Electromagnetic waves induce a current in the earth's surface due to which it gets slowed down(attenuated) due to absorption near the earth causing the waves to tilt downward and hence follow earth's curvature
- In this range, the electromagnetic wave is scattered by the atmosphere layers in such a way that they do not penetrate the upper atmosphere
- Diffraction causes waves to bend
- More the frequency, more is the attenuation
- Distance travelled is short
- For high frequency waves, ionospheric propagation is used

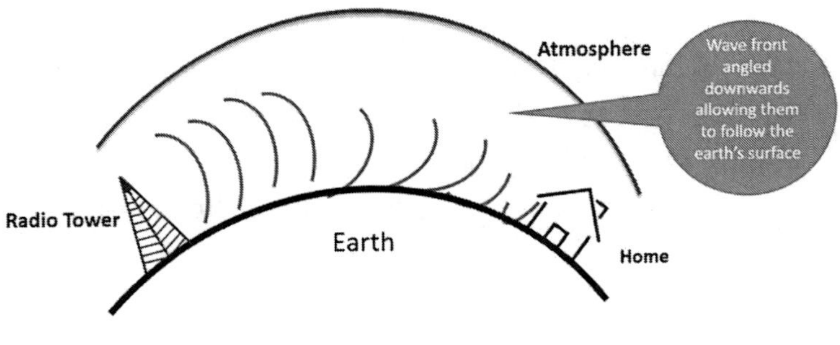

Factors effecting distance travelled by ground waves

- Power of Transmitter; If the Power is high, distance covered will be more.
- Frequency of Transmitter; More the frequency, lesser the distance travelled
  - o long distance communication using frequencies below 3 MHz (the earth behaves as a conductor for all frequencies below 5 MHz)
  - o short distance communications using frequencies between 3 and 30 MHz
- Condition of earth's surface
  - o Ground conductivity, terrain roughness and the dielectric constant all affect the signal attenuation
  - o ground penetration varies, becoming greater at lower frequencies
  - o If there are obstructs like buildings, hills, a distance travelled will be less due to attenuation

## Sky Wave Propagation or Ionospheric Propagation (2-30 MHz)

Atmosphere consists of different layers. When cosmic rays from sun fall on atmospheric layers, atoms get ionised. At higher layers, the density is less; less ions are produced. Solar radiation penetrates that layer and reaches the middle layer. Middle layer is denser; more ions are produced. Radiation penetrates further; but less radiation reaches the lower layers; though denser less ions are produced. Middle layer maximum ionisation; known as Ionosphere. High frequency waves (3-32 MHz) when reaches this layer and gets reflected to the ground station hence covering large distance. Phenomena are similar to total internal reflection in optics. A signal from antenna at a station on earth strikes the ionosphere; It is reflected back down the earth. Sky wave signal bounces back and forth between the ionosphere and earth's surface and can travel along a long distance across the globe. Used in Amateur radio, CB radio, International Broadcast, BBC, voice of America.

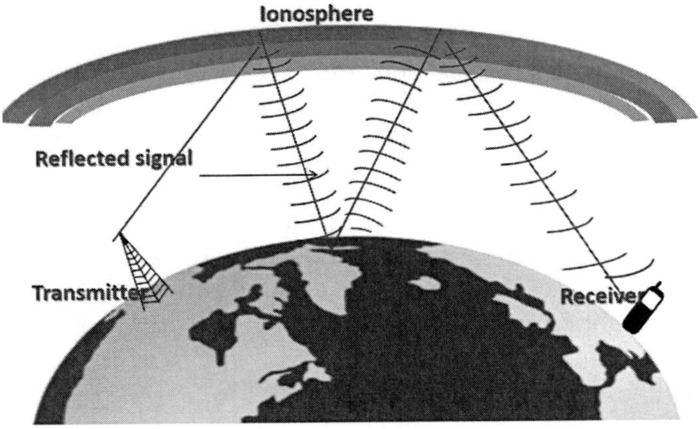


Figure- 4 Sky Waves Propagation

Distance travelled by sky waves depends on

- Angle of Radiation: If the angle of radiation from the transmitting antenna towards sky is travelling short distance and if angle of radiation is more, sky waves will travel longer distance.
- Strength of Ionosphere: If the strength of Ionosphere is more, attenuation of waves will be less, (this is how sky wave broadcast travel all over the world) they will travel longer distance and if strength is less, it will offer more attenuation to the sky waves hence they will be received at a shorter distance.
- Power of transmitter: More the power, more the distance from transmitter to receiver.
- Level of ionization: It depends on Level of radiation which decreases with altitude, time of day of year and season. As per the sun condition, the ionization increases and decreases. More the ionization, more the reflection. Therefore, the propagation also changes as per day and night. Frequencies below 10MHz propagation effectively by sky wave at night whereas frequencies above 10 MHz propagate effectively during the day.
- Line-of-sight or Space wave The communication follows straight line-of-sight at frequency higher than 40 MHz At this frequency, the signal is not reflected by ionosphere and it penetrates the ionosphere and escape. In satellite communication, this signal straight reaches

the satellite overhead and back to the ground station at Earth penetrating layers of atmosphere. For ground based communications, this propagation follows line-of-sight i.e. the transmitter and receiver should be in line-of-sight. Therefore the space wave propagation is also called as line of sight propagation.

Example satellite communication, FM broadcast, two-way radio, AM aircraft communication, Aircraft navigation aids, cellular phone, radar, microwaves links, cordless phones, all Infrared and optical communications use Line-of-sight Propagation or space wave propagation.

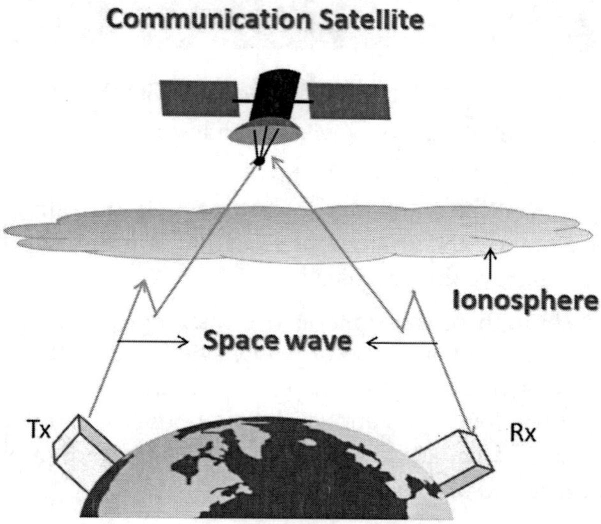


Figure- Space wave Communication

Factors affecting distance travelled by space waves

- Line of sight distance : Distance between transmitting antenna and receiving antenna at which they can see each other, which is also called range of communication
- Curvature of the earth: LOS waves gets blocked by curvature of earth

To increase range, the heights of transmitting and receiving antennas should be increased. The height should be above the horizon If R is radius of the earth, antenna height should be ht = √ ht Maximum LOS distance= √ ht +√ hr Where ht and hr are heights of sender and receiving antennas.

# Phenomena affecting signal propagation

When a signal travels in free space, it behaves like light. It travels straight from sender to receiver. But in real scenario signal in mobile communication, Wi-Fi and other networks have to come across various objects in atmosphere, environment and other surroundings like mountains, buildings, furniture, rain water molecules of atmosphere layer and many others. These objects pose different effects on the propagation of signals namely

- Reflection
- Diffraction
- Refraction
- Scattering

As a result the signal received will differ from signal transmitted due to the various transmission impairments caused by these phenomena. For analog signals these impairments degrade quality of signals whereas in digital data they may cause bit errors like a bit changed to 1 or vice versa.

## Reflection

When the propagating signal falls on an object whose size is large as compared to wavelength of the signal it is reflected. For example, surface of earth buildings and walls offer reflection to the propagating signal For example: A GSM signal whose frequency = 900 MHz; $\lambda$ = 33 cm; Signal will be reflected by objects of size > 10m. Reflection depends on dielectric properties of objects. If the object is a perfect dielectric some energy will be reflected back, some will be transmitted and there will be no absorption of energy. If the object is perfect conductor all the energy of the signal will be reflected back. For long distance sea is one of the best reflecting surfaces. Desert areas are poor reflector. For short range common's buildings, metallic surface offer reflects. The strength of reflected signal will be decreased if the medium or the object offers absorption of signal. Though reflection degrades the signal quality, it also helps the signal to reach the points where LOS transmitter cannot reach.

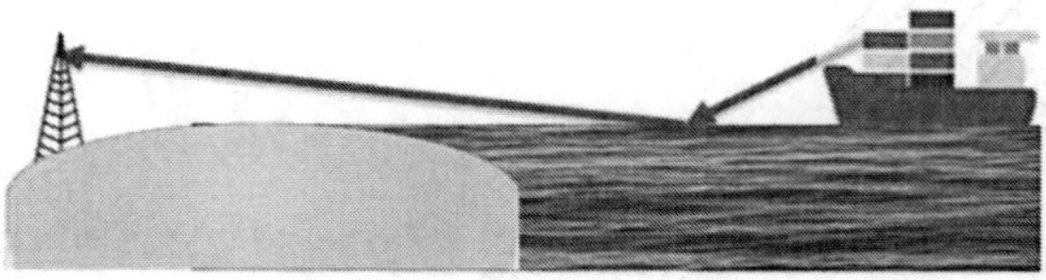

Figure 6 Reflection from sea

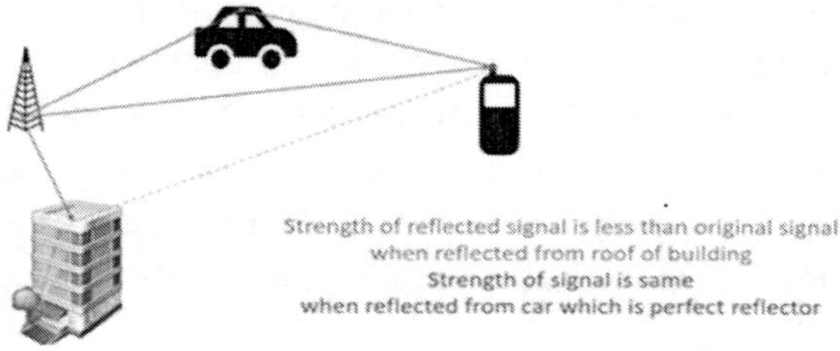


Figure- Reflection from different object

## Diffraction

Diffraction of radio waves is caused by obstructs or surfaces that has sharp irregularities (edge). Corners of buildings, furniture, back of trees all offer diffraction. The phenomena causes bending of radio wave as it passes the edge of the object. It is due to redistribution of energy within a wave front when it passes near the edge of an opaque object and allows radio wave to propagate around corners. Due to this effect, radio signals propagate around the surface of the earth, beyond the horizon and behind the obstructions. Diffraction results in a change of direction of part of the wave energy from the normal line-of-sight path. This change makes it possible to receive energy around the edges of an obstacle. The lower the frequency or the longer the wavelength, the greater the bending of the wave. Radio waves are more readily diffracted than light waves.

Figure- Diffraction

## Scattering

When the dimensions of the objects in the path of radio wave are very small as compared to the wavelength of the waves scattering takes place. Example: forage, street signs, lamp posts offer scattering. The incoming signal is scattered into several weaker outgoing signals.

Figure- 9 Scattering

For optical light of wavelength of order nanometer, water molecules offer scattering. For example: when sunlight fall a water droplet in the cloudlets it gets scattered into

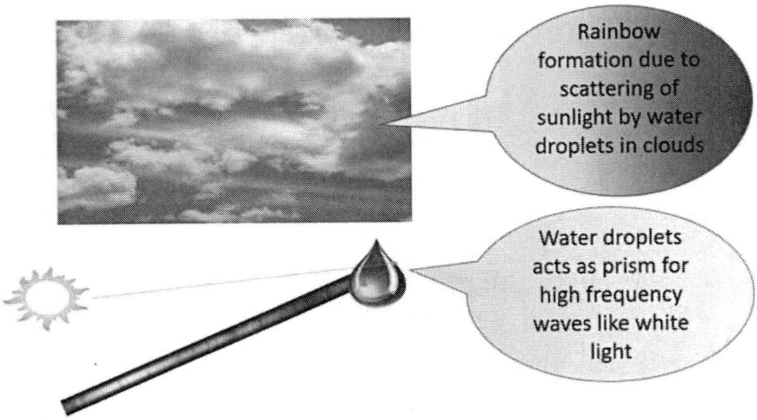


Figure- 10 Scattering

constituent colors – VIBGYOR. Here water droplets act as prism and we see a rainbow.

## Transmission impairments

All these effects have both positive and negative impact on propagation. They help in propagation when there is no clear Line of Sight between sender and receiver. For example: WLAN in office environment, mobile communication in street. at the same time, they also offer some propagation impairments that degrade the quality of signals. All these effects may occur individually or simultaneously and depends on wavelength of wave and also on time. Therefore it is very difficult to interpret or predict the strength of signal at a particular point of time. But it is also necessary to do that in advance to decide the location of antenna, direction of bean and cover age of antenna. Therefore propagation models are used for predicting the average received signal strength at a given distance from transmitter.

Two types of models are used to find the average signal strength over a transmitter received separation distances:

- Large scale propagation model
- Small scale propagation model

If the distance is in hundreds or thousands of meters as in case of mobile physical large scale is used and for short travel distance small scale models are used. The next sections present some transmission impairments and their effects on quality of communication:

## Attenuation

Attenuation is reduction of signal strength during transmitter. Its unit is decibels .It is also known as path loss.

Example If an RF signal transmitted at 10mw encounter a wall which reduces its power to

5mw.

Attenuation =1 =10 * 0.3010=3 decibels

For an ideal isotropic antenna, free space loss is

(Attenuation increase with frequency)

Or

For other antennas, gain of antenna should be considered

Therefore received power depends on

1. Increase frequency of propagation

2. Distance depends between transmission and receiver

3. Depends on gain of transmitting and receiving power in case antenna is not isotropic

## Free Space loss

Free space propagation occurs when the transmitting and receiving antenna are located in an empty environment. The obstacles are absent in such an environment. Radio waves propagate as light does i.e. follow line-of-sight. But even if no matter exists between sender and receiver, the signal experience free space loss. This is the main mode of loss, the strength of the signal falls with increase in distance. This is because for isotropies radiation, there is omni directional propagation i.e. it disperses with distance in the form of a sphere with transmitter as the center. If no obstacles are there, this sphere grows and the sending energy is equally distributed on the surface of the sphere area with radius as the distance from the transmitter equal to $4\pi d^2$ . Therefore, signal strength decreases with square of distance from the transmitter.

Example: If the transmitter sends signal of strength at distance of 100m. Then signal strength at 400m (4 times increase distance)

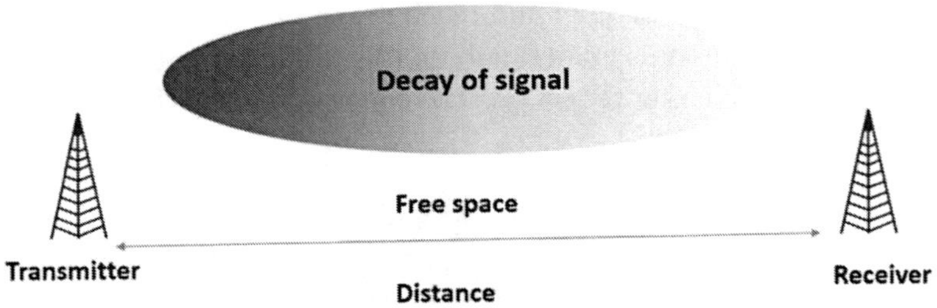


**Attenuation in free space**

Figure- Attenuation in free space

We have discussed that signal strength in free space decreases only as the distance. It behaves very much like light but in real life, the propagation likes in mobile phone, WLAN etc. do not follow line-of-sight. Particularly in urban area there are many objects like buildings, skyscrapers, trees etc. are there. They also affect quality of signal. Besides these atmosphere components likes rain, fog, air, smog influences the propagation. They absorb much of radiation of antenna.

## Blocking or shadowing

Blocking or shadowing is extreme form of attenuation in which the radio signals are blocked by obstacles. The effect is frequency dependent greater the frequency more it behaves likes light gets obstructed even by small obstacle. Lower the frequency, penetration is more. Long waves can even penetrate through the ocean while small waves can blocked even by a tree. Water vapor and oxygen in the atmosphere attenuates signals in the range of 22GHz. Oxygen absorbs signals of 30GHz – 60GHz frequency. Rain and fog attenuates by scattering. Therefore in the areas of precipitates lower frequency bands should be used. Flat surfaces with dimensions larger than a wavelength often reflection whereas rough surface contribute to scattering.

## Multipath Propagation

When the signal travels from sender to receiver through the radio channel, it is reflected, refracted, diffracted or scattered by the objects in the environment like buildings, atmosphere etc in mobile telephony and furniture in WLAN. As a result the signal does not follow line-of-sight and follow different propagation paths in the form of multipath components. Each multipath component travel with different amplitude, delay and phase shifts among each other and reach the receiver. In the figure below it can be seen that signal is taking many paths. One is direct line of sight and others are followed after reflection, diffraction and scattering.

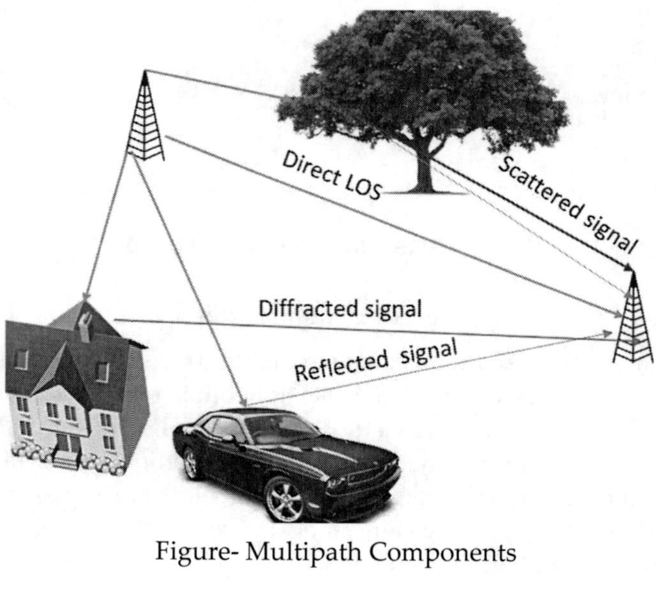


Figure- Multipath Components

## Delay Spread

Due to finite speed, multipath components travelling along different paths with different lengths arrive at receiver at different times. Each component may suffer attenuation and arrive with different power. Some may not be detected at all and rejected as noise. This effect is called delay spread. In cities delay spread is 3 to 12 micro sec. GSM can tolerate upto 16 micro sec delay spread. It can be seen in Fig. that 3 signals have arrived the receiver at different times with different power.

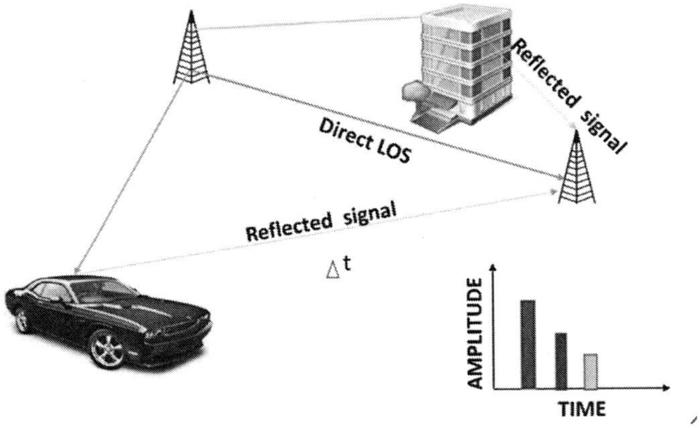


Figure- Delay Spread

### Inter Symbol Interference

Each multipath component arrive the receiver as a sequence of pulses with different time, amplitude and phase called Inter symbol interference. This effect occurs when data is being transmitted. In the Fig. symbols s1, s2, s3, s4 are transmitted. At first path, there is no delay. In second path, there is delay of one symbol period. In there it is of two symbol periods.

**Therefore, at receiver, the symbols interfere as shown in figure.**

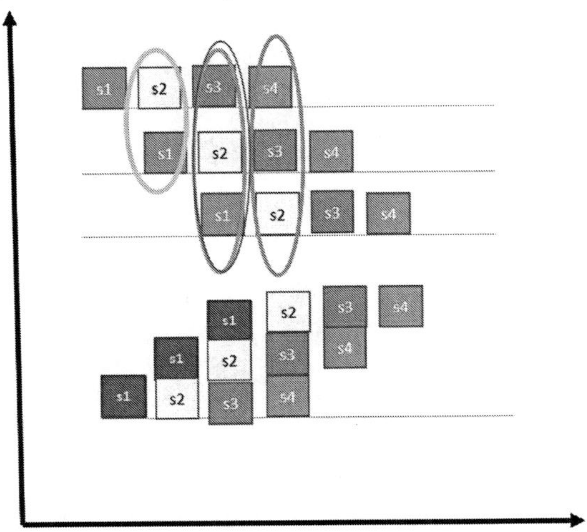

### Interference of multipath componentss

When multipath components with different power arrive at the receiver in phase with each other, they add constructively and phenomena is known as constructive interference. Resultant power is sum of powers of individual components.

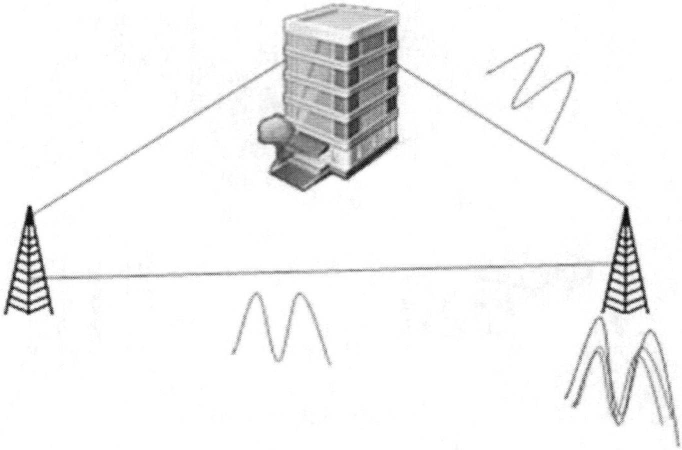

Figure- Constructive interference

When multipath components with different power arrive at the receiver out of phase with each other, they add destructively and phenomena is known as destructive interference. Resultant power is difference of powers of individual components known as fading.

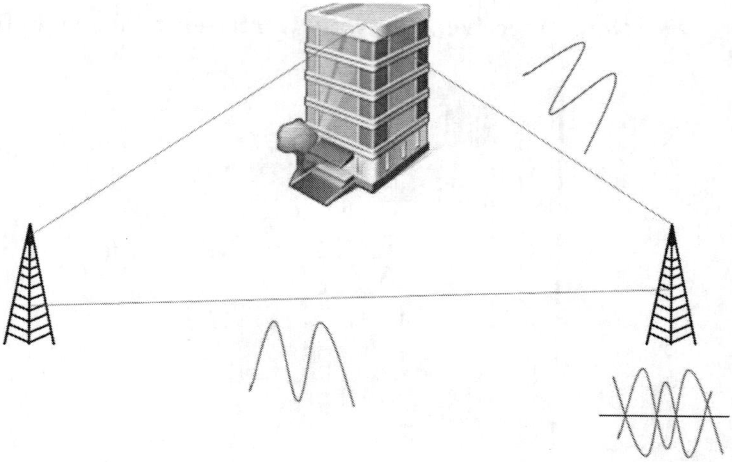

Figure- Destructive Interference

## Solutions for Multipath propagation

MIMO and OFDM are the solution for multipath problem. OFDM signals can be made orthogonal to each other so that they do not interfere. It is used by 802.11n, Wi-Fi, LTE-advanced, Wimax and many more technologies. MIMO is a technology which uses multiple antennas to setup multiple data streams on the same channel to increase its capacity. It takes advantage of multipath propagation in providing additional capacity to the channel.

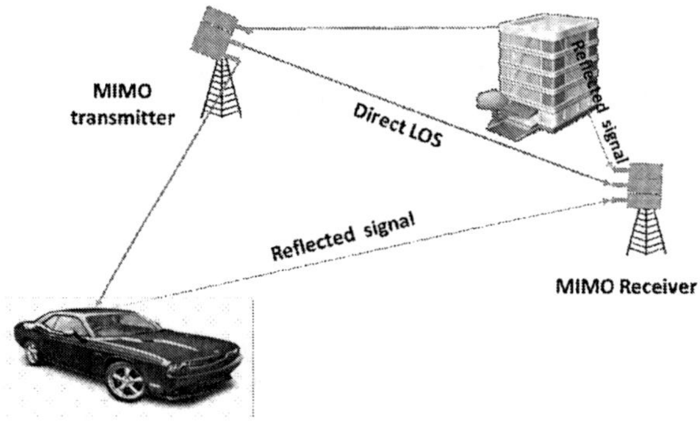


Figure- MIMO

# Log distance path loss model

Log distance path loss model is an extension to the Friis free space model. It is used to predict the propagation loss for a wide range of environments, whereas, the Friis free space model is restricted to unobstructed clear path between the transmitter and the receiver. The model encompasses random shadowing effects due to signal blockage by hills, trees, buildings etc. It is also referred as log normal shadowing model.

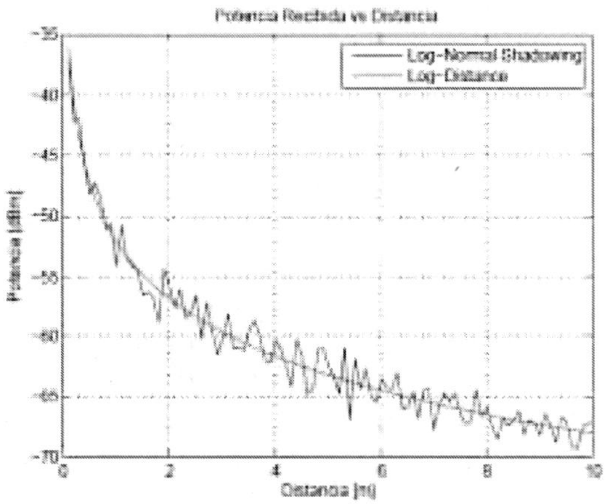

In the far field region of the transmitter, for distances beyond $d_f$, if $P_L(d_0)$ is the path loss at a distance $d_0$ meters from the transmitter, then the path loss at an arbitrary distance $d > d_0$ is given by

$$[P_L(d)]dB = [P_L(d_0)]dB + 10\ n\ log_{10}\left(\frac{d}{d_0}\right) + \chi$$

$$for\ d_f \leq d_0 \leq d \qquad (1)$$

where, $P_L(d)$ is the path loss at an arbitrary distance $d$ meters, $n$ is the path loss exponent that depends on the type of environment, as given in Table below. Also, $X$ is a zero-mean Gaussian distributed random variable with standard deviation $\sigma$ expressed in $dB$, used only when there is a shadowing effect. The reference path loss $P_L(d_0)$, also called close-*in* reference distance, is obtained by using Friis path loss equation

$$P_r(d) = P_t \frac{G_t G_r \lambda^2}{(4\pi)^2 d^n L} \qquad (2)$$

or by field measurements at $d_0$. Typically, $d_0$=*1m* to *10m* for microcell and $d_0$ = *1 Km* for a large cell.

| Environment | Path Loss Exponent (n) |
|---|---|
| Free space | 2 |
| Urban area cellular radio | 2.7 to 3.5 |
| Shadowed urban cellular radio | 3 to 5 |
| Inside a building - line-of-sight | 1.6 to 1.8 |
| Obstructed in building | 4 to 6 |
| Obstructed in factory | 2 to 3 |

*Table 1: Path loss exponent for various environments*

Figure above shows the received signal power when there is no shadowing effect and the case when shadowing exists. The results are generated for an environment with PLE $n$ = 2, frequency of transmission $f$ = *2.4 GHz*, reference distance $d_0$ = *1 m* and standard deviation of the log-normal shadowing $\sigma$ = *2dB*. Results clearly show that the log-normal shadowing introduces randomness in the received signal power, which may put us close to reality.

# MULTIPLE ACCESS SCHEMES

Multiple access schemes are used to allow many mobile users to share simultaneously a finite amount of radio spectrum.

## Multiple Access Techniques

In wireless communication systems, it is often desirable to allow the subscriber to send information simultaneously from the mobile station to the base station while receiving information from the base station to the mobile station.

A cellular system divides any given area into cells where a mobile unit in each cell communicates with a base station. The main aim in the cellular system design is to be able to increase the capacity of the channel, i.e., to handle as many calls as possible in a given bandwidth with a sufficient level of quality of service.

There are several different ways to allow access to the channel. These includes mainly the following –

- Frequency division multiple-access (FDMA)
- Time division multiple-access (TDMA)
- Code division multiple-access (CDMA)
- Space division multiple access (SDMA)

Depending on how the available bandwidth is allocated to the users, these techniques can be classified as narrowband and wideband systems.

## Narrowband Systems

Systems operating with channels substantially narrower than the coherence bandwidth are called as Narrow band systems. Narrow band TDMA allows users to use the same channel but allocates a unique time slot to each user on the channel, thus separating a small number of users in time on a single channel.

## Wideband Systems

In wideband systems, the transmission bandwidth of a single channel is much larger than the coherence bandwidth of the channel. Thus, multipath fading doesn't greatly affect the received signal within a wideband channel, and frequency selective fades occur only in a small fraction of the signal bandwidth.

### Frequency Division Multiple Access (FDMA)

FDMA is the basic technology for advanced mobile phone services. The features of FDMA are as follows.

- FDMA allots a different sub-band of frequency to each different user to access the network.
- If FDMA is not in use, the channel is left idle instead of allotting to the other users.
- FDMA is implemented in Narrowband systems and it is less complex than TDMA.
- Tight filtering is done here to reduce adjacent channel interference.
- The base station BS and mobile station MS, transmit and receive simultaneously and continuously in FDMA.

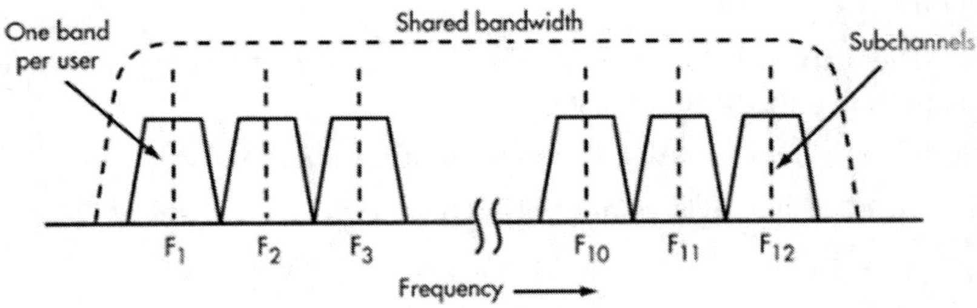


### Time Division Multiple Access (TDMA)

In the cases where continuous transmission is not required, there TDMA is used instead of FDMA. The features of TDMA include the following.

- TDMA shares a single carrier frequency with several users where each users makes use of non-overlapping time slots.
- Data transmission in TDMA is not continuous, but occurs in bursts. Hence handsoff process is simpler.
- TDMA uses different time slots for transmission and reception thus duplexers are not required.
- TDMA has an advantage that is possible to allocate different numbers of time slots per frame to different users.

- Bandwidth can be supplied on demand to different users by concatenating or reassigning time slot based on priority.

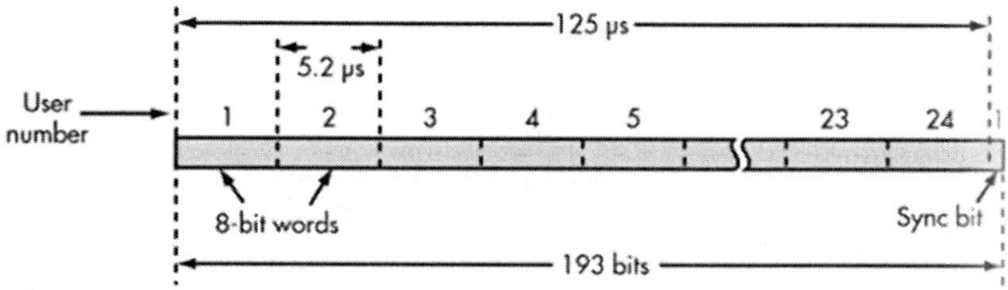


## Code Division Multiple Access (CDMA)

Code division multiple access technique is an example of multiple access where several transmitters use a single channel to send information simultaneously. Its features are as follows.

- In CDMA every user uses the full available spectrum instead of getting allotted by separate frequency.
- CDMA is much recommended for voice and data communications.
- While multiple codes occupy the same channel in CDMA, the users having same code can communicate with each other.
- CDMA offers more air-space capacity than TDMA.
- The hands-off between base stations is very well handled by CDMA.

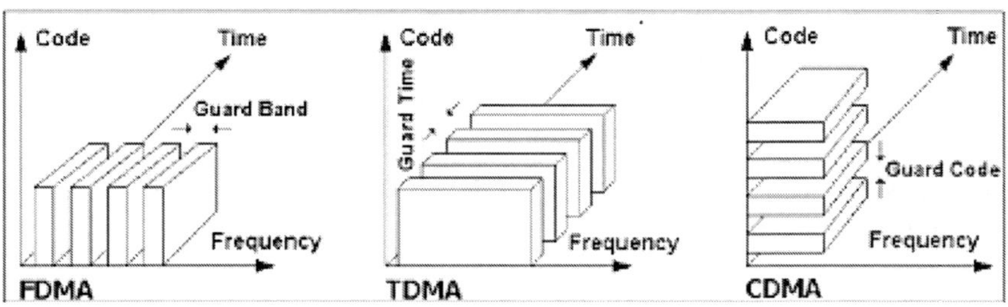


## Space Division Multiple Access (SDMA)

Space division multiple access or spatial division multiple access is a technique which is MIMO (multiple-input multiple-output) architecture and used mostly in wireless and satellite communication. It has the following features.

- All users can communicate at the same time using the same channel.
- SDMA is completely free from interference.
- A single satellite can communicate with more satellites receivers of the same frequency.
- The directional spot-beam antennas are used and hence the base station in SDMA, can track a moving user.
- Controls the radiated energy for each user in space.

### Spread Spectrum Multiple Access

Spread spectrum multiple access (SSMA) uses signals which have a transmission bandwidth whose magnitude is greater than the minimum required RF bandwidth.

There are two main types of spread spectrum multiple access techniques –

- Frequency hopped spread spectrum (FHSS)
- Direct sequence spread spectrum (DSSS)

### Frequency Hopped Spread Spectrum (FHSS)

This is a digital multiple access system in which the carrier frequencies of the individual users are varied in a pseudo random fashion within a wideband channel. The digital data is broken into uniform sized bursts which is then transmitted on different carrier frequencies.

### Direct Sequence Spread Spectrum (DSSS)

This is the most commonly used technology for CDMA. In DS-SS, the message signal is multiplied by a Pseudo Random Noise Code. Each user is given his own code word which is orthogonal to the codes of other users and in order to detect the user, the receiver must know the code word used by the transmitter.

The combinational sequences called as hybrid are also used as another type of spread spectrum. Time hopping is also another type which is rarely mentioned.

Since many users can share the same spread spectrum bandwidth without interfering with one another, spread spectrum systems become bandwidth efficient in a multiple user environment.

## Difference between FDMA, TDMA and CDMA

**1. Frequency Division Multiple Access (FDMA) :** FDMA is a type of channelization protocol. In this bandwidth is divided into various frequency bands. Each station is allocated with band to send data and that band is reserved for particular station for all the time which is as follows :

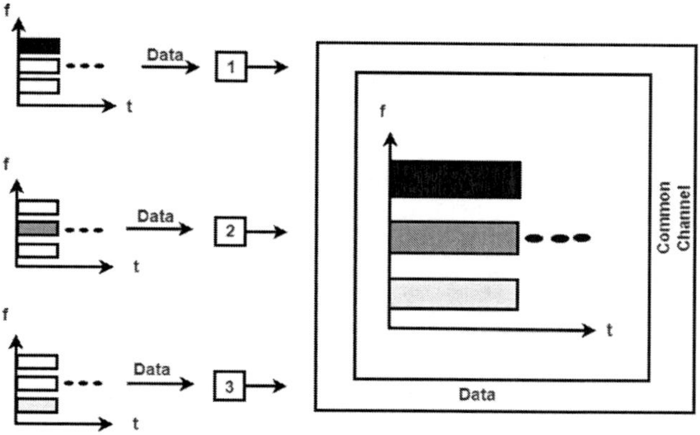


*Figure – FDMA*

The frequency bands of different stations are separated by small band of unused frequency and that unused frequency bands are called as guard bands that prevents the interference of stations. It is like access method in data link layer in which data link layer at each station tells its physical layer to make a band pass signal from the data passed to it. The signal is created in the allocated band and there is no physical multiplexer at the physical layer.

**2. Time Division Multiple Access (TDMA) :** TDMA is the channelization protocol in which bandwidth of channel is divided into various stations on the time basis. There is a time slot given to each station, the station can transmit data during that time slot only which is as follows :

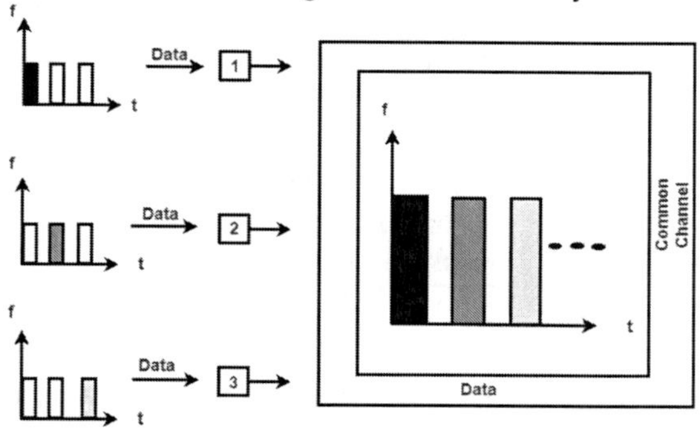


*Figure – TDMA*

Each station must aware of its beginning of time slot and the location of the time slot. TDMA requires synchronization between different stations. It is type of access method in the data link layer. At each station data link layer tells the station to use the allocated time slot.

**3. Code Division Multiple Access (CDMA) :** In CDMA, all the stations can transmit data simultaneously. It allows each station to transmit data over the entire frequency all the time. Multiple simultaneous transmissions are separated by unique code sequence. Each user is assigned with a unique code sequence.

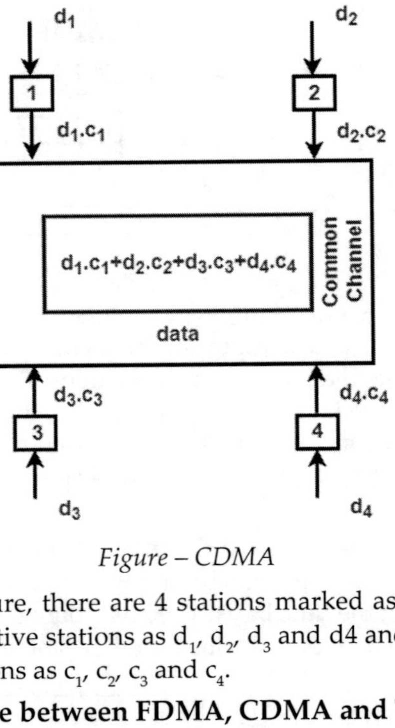


*Figure – CDMA*

In the above figure, there are 4 stations marked as 1, 2, 3 and 4. Data assigned with respective stations as $d_1$, $d_2$, $d_3$ and d4 and the code assigned with respective stations as $c_1$, $c_2$, $c_3$ and $c_4$.

**Difference between FDMA, CDMA and TDMA :**

| FDMA | TDMA | CDMA |
|---|---|---|
| FDMA stands for Frequency Division Multiple Access. | TDMA stands for Time Division Multiple Access. | CDMA stands for Code Division Multiple Access. |
| In this, sharing of bandwidth among different stations takes place. | In this, only the sharing of time of satellite transponder takes place. | In this, there is sharing of both i.e. bandwidth and time among different stations takes place. |

| FDMA | TDMA | CDMA |
|---|---|---|
| There is no need of any codeword. | There is no need of any codeword. | Codeword is necessary. |
| In this, there is only need of guard bands between the adjacent channels are necessary. | In this, guard time of the adjacent slots are necessary. | In this, both guard bands and guard time are necessary. |
| Synchronization is not required. | Synchronization is required. | Synchronization is not required. |
| The rate of data is low. | The rate of data is medium. | The rate of data is high. |
| Mode of data transfer is continuous signal. | Mode of data transfer is signal in bursts. | Mode of data transfer is digital signal. |
| It is little flexible. | It is moderate flexible. | It is highly flexible. |

## GSM in Wireless Communication

GSM (Global System for Mobile Communication) is a digital mobile network commonly utilized by mobile phone users in Europe and around the world. GSM, the most popular of the three digital wireless telephony systems (TDMA, GSM, and CDMA), is a version of time division multiple access (TDMA). GSM converts and compresses data before sending it along a channel with two other streams of user data, each with its time slot.

### What is GSM?

GSM stands for Global System for Mobile Communication. GSM is an open and digital cellular technology used for mobile communication. It uses 4 different frequency bands 850 MHz, 900 MHz, 1800 MHz, and 1900 MHz. It uses the combination of FDMA and TDMA.

### GSM is having 4 different sizes of cells

- **Macro:** In this size of the cell, a Base Station antenna is installed.
- **Micro:** In this size of cell, antenna height is less than the average roof level.
- **Pico:** Small cells' diameter of a few meters.
- **Umbrella:** It covers the shadowed (Fills the gaps between cells) regions.

### Features of GSM

- Supports international roaming
- Clear voice clarity
- Ability to support multiple handheld devices.
- Spectral / frequency efficiency
- Low powered handheld devices.
- Ease of accessing network
- International ISDN compatibility.
- Low service cost.
- New features and services.

### The Architecture of GSM

- **BSS :** BSS stands for Base Station Subsystem. BSS handles traffic and signaling between a mobile phone and the network switching subsystem. BSS having two components **BTS** and **BSC.**
- **NSS :** NSS stands for Network and Switching Subsystem. NSS is the core network of GSM. That carried out call and mobility management functions for mobile phone present in network. NSS have different components like **VLR, HLR** and **EIR.**
- **OSS :** OSS stands for Operating Subsystem. OSS is a functional entity which the network operator monitor and control the system. **OMC** is the part of OSS. Purpose of OSS is to offer the customer cost-effective support for all GSM related maintenance services.

### The Architecture of GSM

The GSM architecture is made up of three central systems. The following are the primary components of the GSM architecture:

- The network switching system (NSS)
- The mobile station (MS)
- The base station system (BSS)
- The operations and support system (OSS)

### 1. The network switching system (NSS)

NSS is a GSM element that provides flow management and call processing for mobile devices moving between base stations. The switching system consists of the functional units listed below.

- **Mobile Services Switching Center (MSC)**: Mobile Switching Center is integral to the GSM network architecture's central network space. The MSC supports call switching across cellular phones and other fixed or mobile network users. It also monitors cellular services, including registration, location updates, and call forwarding to a roaming user.
- **Home Location Register (HLR): I**t is a set of data items used for storing and managing subscriptions. It provides data for each consumer as well as their last known position. The HLR is regarded as the most significant database because it preserves enduring records about users. When a person purchases a membership from one of the operators, they are enlisted in that operator's HLR.
- **Visitor Location Register (VLR):** VLR is a database that provides subscriber information necessary for the MSC to service passengers. This includes a short-term version of most of the data stored in the HLR. The visitor location register can also be run as a standalone program, but it is usually implemented as a component of the MSC.
- **Equipment Identity Register (EIR):** It is the component that determines if one can use particular mobile equipment on the system. This consists of a list of every functioning mobile device on the system, with each mobile device recognized by its own International Mobile Equipment Identity (IMEI) number.
- **Authentication Center (AuC):** The AUC is a unit that offers verification and encryption factors to ensure the user's identity and the privacy of every call. The verification center is a secure file that contains the user's private key in the SIM card. The AUC shields network operators from various types of fraud prevalent in the modern-day cellular world.

## 2. The mobile station (MS)

The mobile station is a cell phone with a display, digital signal processor, and radio transceiver regulated by a SIM card that functions on a system. Hardware and the SIM card are the two most essential elements of the MS. The MS (Mobile stations) is most widely recognized by cell phones, which are components of a GSM mobile communications network that the operator monitors and works.

Currently, their size has shrunk dramatically while their capabilities have skyrocketed. Additionally, the time between charges has been significantly improved.

### 3. The base station system (BSS)

It serves as a connection between the network subsystem and the mobile station. It consists of two parts:

- **The Base Transceiver Station (BTS):** The BTS is responsible for radio connection protocols with the MS and contains the cell's radio transceivers. Companies may implement a significant number of BTSs in a big metropolitan area. Each network cell has transceivers and antennas that make up the BTS. Based on the cell's consumer density, every BTS includes anywhere from one to sixteen transceivers.
- **The Base Station Controller (BSC):** The BSC is responsible for managing the radio resources of one or more BTS(s). This manages radio channel configuration and handovers. The BSC serves as the link seen between mobile and MSC. It allocates and emits MS frequency bands and time slots. Additionally, the BSC is responsible for intercell handover and transmits the BSS and MS power within its jurisdiction.

### 4. The operations and support system (OSS)

The operation support system (OSS) is a part of the overall GSM network design. This is linked to the NSS and BSC components. The OSS primarily manages the GSM network and BSS traffic load. As the number of BS increases due to customer population scaling, a few maintenance duties are shifted to the base transceiver stations, lowering the system's financial responsibility. The essential purpose of OSS is to have a network synopsis and assist various services and maintenance organizations with their routine maintenance arrangements.

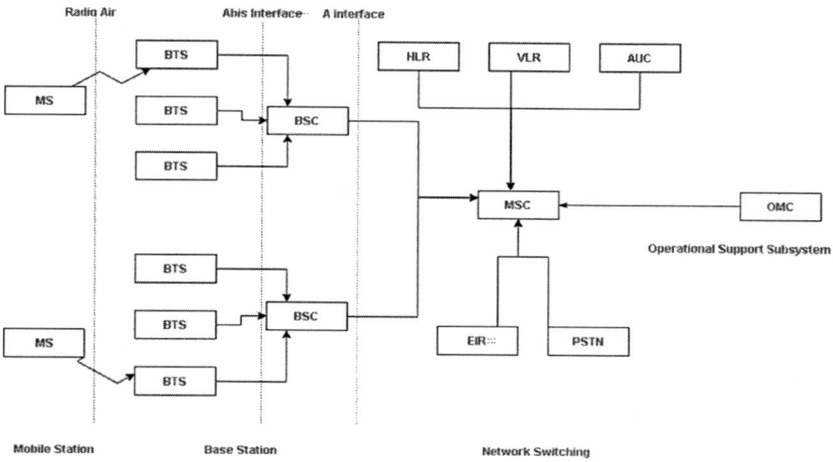


Suppose there are 3 Mobile stations which are connected with the tower and that tower is connected to BTS through TRX, then further connected to BSC and MSC. Let's understand the functionality of different components.

- **MS :** MS stands for Mobile System. MS comprises user equipment and software needed for communication with a mobile network. Mobile Station (MS) = Mobile Equipment (ME) + Subscriber Identity Module (SIM). Now, these mobile stations are connected to tower and that tower connected with BTS through TRX. TRX is a transceiver which comprises transmitter and receiver. Transceiver has two performance of sending and receiving.
- **BTS :** BTS stands for Base Transceiver Station which facilitates wireless communication between user equipment and a network. Every tower has BTS.
- **BSC :** BSC stands for Base Station Controller. BSC has multiple BTS. You can consider the BSC as a local exchange of your area which has multiple towers and multiple towers have BTS.
- **MSC :** MSC stands for Mobile Switching Center. MSC is associated with communication switching functions such as call setup, call release and routing. Call tracing, call forwarding all functions are performed at the MSC level. MSC is having further components like VLR, HLR, AUC, EIR and PSTN.

- **VLR :** VLR stands for Visitor Location Register. VLR is a database which contains the exact location of all mobile subscribers currently present in the service area of MSC. If you are going from one state to another state then your entry is marked into the database of VLR.
- **HLR :** HLR stands for Home Location Register. HLR is a database containing pertinent data regarding subscribers authorized to use a GSM network. If you purchase SIM card from in the HLR. HLR is like a home which contains all data like your ID proof, which plan you are taking, which caller tune you are using etc.
- **AUC :** AUC stands for Authentication Center. AUC authenticates the mobile subscriber that wants to connect in the network.
- **EIR :** EIR stands for Equipment Identity Register. EIR is a database that keeps the record of all allowed or banned in the network. If you are banned in the network then you can't enter the network, and you can't make the calls.
- **PSTN :** PSTN stands for Public Switched Telephone Network. PSTN connects with MSC. PSTN originally a network of fixed line analog telephone systems. Now almost entirely digital in its core network and includes mobile and other networks as well as fixed telephones. The earlier landline phones which places at our home is nothing but PSTN.

- **OMC** : OMC stands for Operation Maintenance Center. OMC monitor and maintain the performance of each MS, BSC and MSC within a GSM system.

Three subsystem BSS, NSS and OSS are connected with each other via some interfaces. Total three interfaces are there:

- **Air Interface :** Air interface is also known as UM interface. Interface between MS and BTS is called as UM interface because it is mobile analog to the U interface of ISDN.

- **Abis Interface :** It is a BSS internal interface linking with BTS and BSC.
- **A interface :** It provides communication between BSS and MSC.

### How Does GSM Work?

GSM is a globally recognised digital cellular communication protocol. The GSM standard was developed by the European Telecommunications Standards Institute to describe the procedures for second-generation digital mobile networks, such as those used by mobile phones. It is a broad-area communications technology programme that uses digital radio channelling to provide audio, information, and multimedia communication systems. Every GSM radio channel is 200 kHz broad and is further divided into frames of eight time slots. The GSM system consists of mobile stations, base stations, and interweaving switching systems. The GSM programme allows 8 to 16 audio users to share a single radio channel, and each radio transmission station can have numerous radio channels. Because of its simplicity, cost, and accessibility, GSM is now the most often utilised network technology in the Internet of Things (IoT).

### The Working of a GSM Network

GSM is a digital cellular communication standard that is universally accepted. The European Telecommunications Standards Institute created the GSM standard to define the procedures for second-generation digital mobile networks that are used by devices such as mobile phones. It is a wide-area communications technology program that utilizes digital radio channeling to bring forth audio, information, and multimedia communication systems.

GSM is a mobile network and not a computer networkOpens a new window – this implies that devices interact with it by looking for nearby cells. GSM, including other technological advances, has influenced the evolution of mobile wireless telecommunication services. A GSM system manages communication between mobile stations, base stations, and switching systems.

Every GSM radio channel is 200 kHz wide and is additionally divided into frames of 8-time slots. The global system for mobile communication (GSM) was first known as Groupe Special Mobile, which is the reason for the acronym. The GSM system comprises mobile stations, base stations, and intertwining switching systems.

The GSM program enables 8 to 16 audio users to share every radio channel, and every radio transmission location may have multiple radio channels. Because of its simplicity, affordability, and accessibility, GSM is presently the most commonly used network technology in the Internet of Things (IoT)Opens a new window  applications.

However, this is likely to change in the coming years. Various programs have been designed without the advantage of standardized provisions all through the transformation of mobile telecommunication services.

This significantly created many issues tied directly to consistency as digital radio technology advanced. The global system for mobile communication is designed to address these issues. GSM accounts for about 70% of the world's digital cellular services. GSM automates and encodes the information before transmitting it via a channel including three distinct streams of user information inside each time slot. For the vast majority of the world, it is also the leading 2G digital cell phone standard. It governs how cell phones interact with the land-based tower system.

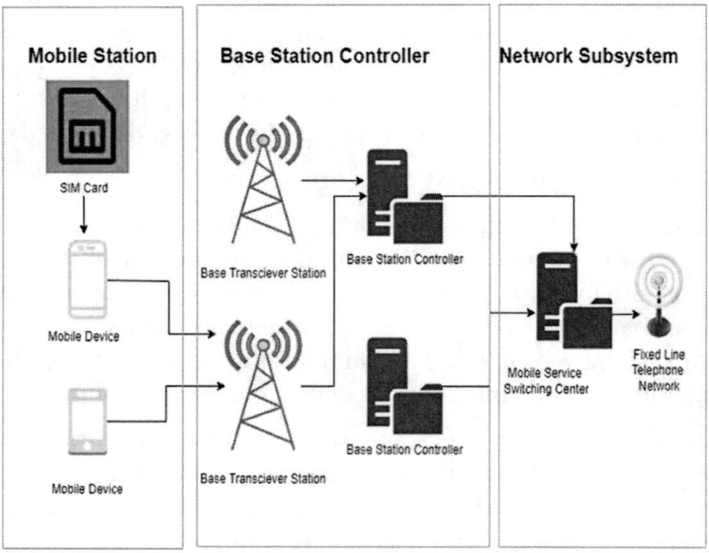


Fig GSM

## Services of GSM

- **Bearer services/ data services:** GSM specifies different mechanism for data transmission, The original GSM allowing for data rates of up to 9600 bits/s. Bearer services permit transparent or non transparent data transmission.
    - **Transparent bearer services:** Transparent bearer services only use the physical layer to transmit

data. Data transmission has a constant delay at throughput if no transmission error occurs.

  - **Non-transparent bearer services:** Non-transparent bearer services use protocols of layer two and three two three to implement error correction and flow control.(data link layer and network layer).

- **Tele services**: Tele services are nothing but we use now as at also.Video calls.
  - Video text and face emoji.
  - Short text message(SMS).
- **Supplementary services:** Supplementary services it means advanced services.
  - Conference calls.
  - Call waiting.
  - Call forwarding.
- **GSM security:**GSM offers several security using confidential information stored in the AUC and in the individual SIM. The SIM stores personal secret data and is protected with a pin against unauthorized use.

## Applications of GSM

- Mobile Telephony
- VoIP Integration
- SMS (Short Message Service)
- Mobile Banking
- Smart Home Systems
- Telemedicine
- Surveillance Systems
- Alarm System
- Cell Broadcasting

## Advantages of GSM

- **Compatibility:** GSM is widely used around the world, so it is compatible with many different networks and devices.

- **Security:** GSM offers enhanced security features such as authentication, encryption and confidentiality, which helps to protect the user's privacy and data.
- **Efficient use of bandwidth:** GSM uses a time-division multiplexing (TDM) technique which enables many users to share the same frequency channel at different times, making it an efficient use of the available bandwidth.
- **Roaming:** GSM allows users to roam internationally and use their mobile phones in other countries that use the same GSM standard.
- **Wide range of features:** GSM supports a wide range of features, including call forwarding, call waiting, voicemail, conference calling, and more.

### Disadvantages of GSM

- **Limited coverage:** GSM networks may have limited coverage in some remote areas, which can make it difficult for users to make calls or access the internet.
- **Network congestion:** GSM networks may become congested during peak hours, which can lead to dropped calls or poor call quality.
- **Security vulnerabilities:** Although GSM offers enhanced security features, it is still vulnerable to certain types of attacks, such as eavesdropping and spoofing.
- **Data transfer speed:** GSM networks offer relatively slow data transfer speeds compared to newer technologies such as 3G and 4G.
- **Limited capacity:** GSM networks have a limited capacity for handling large volumes of data, which can be a disadvantage for users who require high-speed internet access or other data-intensive applications.

### What are GSM vs. CDMA vs. LTE differences?

GSM, CDMA, and LTE (long-term evolution) cellular-wireless communications differ significantly in terms of technology and business objectives. GSM is the oldest of the three. GSM was developed and adopted as a standard in Europe, and it encoded and decoded data using processor/

chip technology available at the time while CDMA is basically a channel access method and is also an example of multiple access. Multiple access basically means that information by several transmitters can be sent simultaneously onto a single communication channel and LTE provides fast peak data transmission, with initial speeds of up to 100 Mbps downstream and 30 Mbps upstream. Due to its high speeds, low latency, and improved capabilities, LTE enables users to enjoy faster data transfer rates, seamless browsing, and enhanced user experience.

### Subscriber Identity Module (SIM)

SIM card as the name suggests is integrated circuit that stores identity of users on mobile telephony devices. SIM card is generally smaller and rectangular, about 25mm by 15mm, and notched at one corner. This feature ensures handy, fail-safe way of inserting the card correctly into the slot in mobile phone. SIM cards can also be used in satellite phones, smart watches etc. SIM cards can be transferred between different phones. SIM cards consist of unique serial number, international mobile subscribe identity (IMSI) number, ciphering information, information related to local network. SIM cards are used in GSM mobile phones while CDMA phones uses Reusable Identity Module (RUIM). It also has limited memory which can be used for storing contacts of person using that SIM card. For example when we save new contact in mobile phone, we are given to options that whether we want to save it in sim card or in phone storage and when we select the option of sim card the contact gets saved in SIM card. This proves that SIM card is portable smart card which is movable among different phones. **SIM Memory Structure :** The SIM memory structure is said to be consist of directories which are analogous to the directories of traditional hard disk drive.The file system consist of Master File(MF), directory file(DF) and elementary file(EF).

1. **Master File** – Master file is root of file system.It is analogous to root directory like in any of operating systems.It may or may not elementary files(EF). It may contain one or many dedicated files(DF).
2. **Dedicated File** – It is sub directory. It can be identified by two byte identifier. The phonebook EF falls under this category.
3. **Elementary File** – These files lie below dedicated files in memory hierarchy system.These files contain actual data. It would not be wrong to say that EF represents leaf node of the system.

**Advantages of Subscriber Identity Module :**

- This tiny card consist of your phone number, rate plan, your contact list and several other things.
- It is portable chip which you can move from one cell phone to other.
- The data on SIM card is protected by PIN lock which is usually set by manufacturer. Although you can change this PIN.

**Disadvantages of Subscriber Identity Module :**

- As the SIM card is portable so if anytime you lose it you lose your personal information like phone number, contact list and several others.
- If you are using dual SIM handset it is possible that phone battery may drain lot.

## Important Questions on Mobile Computing

### 1) What is Mobile Computing? Or What do you understand by Mobile Computing?

Mobile computing is a technology that allows transmission of data, voice and video via a computer or any other wireless-enabled device / mobile device such as a handheld computer, laptop, smartphones, PDA etc. without having to be connected to a fixed physical wire.

Mobile Computing is the process of computation on a mobile device. In this technology, a set of distributed computing systems or service providers and servers participate, connect, and synchronize through mobile communication protocols.

### 2) What are the basic building blocks for Mobile Computing?

Mobile Computing requires portable computing devices such as laptops, smartphones, handheld computers etc. along with mobile communication technologies to enable users to access the Internet and mobile data on their home or work computers from around the world.

### 3) What do you understand by Mobility in Mobile Computing?

Mobility is the ability to move freely. In Mobile Computing, users and devices both have to move on different geographical locations and different networks. Mobility is an essential aspect of Mobile computing.

### 4) What are the two different types of mobility in Mobile Computing?

There are two types of mobility used in Mobile Computing:

**User Mobility:** It is used to specify a user who has access to the same or similar telecommunication services at different places. In User Mobility, a user can move between:

- Different geographical locations.
- Different networks.
- Different communication devices.
- And different applications.

**Device Portability:** It specifies the mobility of the device. There are many mechanisms in the network and inside the device that ensure that communication is still possible while the device is moving.

In Device Portability, the device can move between:

- Different geographical locations.
- Different networks.

### 5) What are the different types of applications in which Mobile Computing is used?

Mobile Computing is used in the following types of applications:

- Mobile and wireless devices
- Location dependent services
- Modern Vehicles
- Medical Emergencies
- Replacement of wired networks
- Infotainment Services
- Business etc.

### 6) What are the main obstacles that occurred in mobile communications?

Following is a list of the biggest obstacles that occurred in mobile communications:

- Interference in data transmission.
- Low Bandwidth during communication.
- Regulations and spectrum.

- High delays, large delay variation.
- Vulnerable to security issues and threats, simpler to attack.
- Shared Medium Adhoc-networks

**7) What is the full form of a SIM card?**

SIM is an acronym that stands for Subscriber Identity Module. It is a small plastic card that contains some unique information.

**8) What information is stored on a SIM card?**

SIM has an ID number or IMSI, which stands for International Mobile Subscriber Identity. This number is used to identify the phone number to the network. SIM basically specifies our phone number but can also store contact information, telephone numbers, SMS messages, billing information and data usage etc.

Along with these features, a SIM card contains the following information:

- The Card type, serial no, and the list of subscribed services.
- Personal Identity Number(PIN)
- Pin Unlocking Key(PUK)
- An Authentication Key(KI)

**9) What does a SIM do? Or What is the working of a SIM card in Mobile communication?**

SIM or SIM card is a computer chip made of plastic that holds information and facilitates connecting to your network. It stores much technical information about network and communication. It facilitates you to make calls, send SMS, and connect to mobile internet services like 3G and 4G. It can store some data such as messages, contacts and emails, etc. You can transfer your contact to SIM in emergency conditions. It is also transferable.

**10) What are the different advantages of Mobile Computing?**

Following is the list of advantages of Mobile Computing:

**Advantages of Mobile Computing**

**Increased Productivity:** Mobile Computing allows users to use mobile devices in various fields and companies. So, it has reduced the time and cost for clients and themselves also that's why productivity is increased.

**Portability:** This is one of the biggest advantages of mobile computing. It allows you to move freely and not restricted to one location to complete your communication or do your jobs. You can even access your mails and do your online assignment while going anywhere.

**Entertainment:** Due to the emerging of mobile devices and smartphones, the entertainment industry is enhanced quickly and bigger than ever. You can use it for entertainment purposes also.

**Cloud Computing:** Mobile Computing provides the service of Cloud computing, which can be used for saving data, documents, audio, video files etc. on an online server and being able to access them anytime and anywhere by having only an internet connection. You can also access these files on several mobile devices or computers simultaneously.

### 11) What are the main disadvantages of Mobile Computing?

Following is the list of main disadvantages of Mobile Computing:

**Disadvantages of Mobile Computing**

**Quality of Connectivity:** The quality of connectivity is one of the biggest disadvantages of Mobile Computing. Mobile devices require either Wi-Fi connectivity or mobile network connectivity such as GPRS, 3G or 4G connectivity. In this case, if you are not near any of these connections, your internet access will be slow, very limited or disconnected sometimes.

**Security Issues and Threats:** Mobile connections are always vulnerable to security issues or threats. Mobile VPNs are not supposed to always be safe, and syncing devices might also lead to security concerns. If you are accessing a Wi-Fi network connection, it can also be risky because hackers may easily hack or bypass WPA and WEP security.

**Power Consumption and Battery Issues:** Mobile devices use detachable and rechargeable batteries as a power source during communication. These batteries do not tend to last long. In emergencies, if you don't have a source of power for charging, then it will undoubtedly be a communication or data transmission failure.

### 12) What do you understand by guard space in Mobile Computing?

The space between the channel interference ranges is known as guard space. Guard spaces are used to avoid frequency band overlapping or adjacent channel overlapping.

### 13) How is guard spaces arranged between users in CDMA systems?

The guard spaces between users in CDMA systems are arranged orthogonally between the spreading codes. The lower the correlation is, the better is the user separation.

**14) What do you understand by multiplexing in Mobile computing?**

**FDM:** The FDM technique separates signals by modulating the data onto different carrier frequencies.

**TDM:** The TDM technique is used to separate signals by interleaving bits one after the other.

**15) What are the three fundamental propagation behaviors depending on their frequency?**

Following are the three fundamental propagation behaviors depending on their frequency:

**Ground wave:** When the frequency is less than 2 MHz. (<2 MHz)

**Sky wave:** When the frequency is varied between 2-30 MHz.

**Line of sight:** When the frequency is more than 30 MHz. (>30 MHz)

**16) What is multipath propagation?/ What do you understand by multipath propagation?**

When the sender emits the radio waves, these radio waves can travel along a straight line, or they may be reflected at a large building, or scattered at smaller obstacles. Such transmission of data from sender to receiver is called multipath propagation.

**17) What are the different basic schemes of analog modulations in mobile computing?**

The three basic schemes analog modulations in mobile computing are:

- Amplitude modulation (AM)
- Frequency modulation (FM)
- Phase modulation (PM)

**18) What are the Phase Lock Loop (PLL) in Mobile computing and its use?**

The Phase Lock Loop or phase-locked loop (PLL) is a control system that is used to generate an output signal whose phase is related to the input signal's phase. There are several different types of a phase-locked loop.

The simplest PLL is an electronic circuit that contains a variable frequency oscillator and a phase detector in a feedback loop.

The phase-locked loop is used to synchronize the frequency and phase of the receiver with the transmitter.

### 19) What is the hopping sequence in Mobile Communication?

The hopping sequence is a pattern of channel usage used in Frequency Hopping Spread Spectrum (FHSS).

### 20) What do you understand by dwell time?

The dwell time is the total amount of time spent on a channel with a specific frequency.

### 21) What are the advantages of a cellular system?

The advantages of a cellular system are:

- Higher capacity
- Consume less transmission power
- Local interference only
- Robustness

### 22) What is the difference between borrowing channel allocation (BCA) and fixed channel allocation (FCA)?

In borrowing channel allocation (BCA), when the traffic load varies in adjacent cells, the cell with heavy load borrows frequencies from the neighboring cell. This scheme is called borrowing channel allocation or BCA.

While in the fixed channel allocation, the frequencies assigned to cell clusters and cells respectively are fixed. This technique is called fixed channel allocation or FCA.

### 23) What is the responsibility of an authentication center?

The responsibility of the Authentication center (AuC) is protecting user identity and data transmission. The AuC uses an algorithm for authentication as well as the keys for encryption. It also generates the values needed for user authentication in the HLR.

### 24) What do you understand by burst and normal burst?

In mobile communication, the data is transmitted in small portions. These small portions are called bursts, and the portions used for data transmission inside a time slot is called normal bursts.

### 25) What is traffic multiframe and control multiframe?

**Traffic multiframe:** The Traffic Channel frames are organized into multiframes that contain 26 bursts and takes 120 ms. 24 bursts are used in a traffic multiframe.

**Control multiframe:** The combination of 51 TDMA frames which contains data for the other logical channels, is called control multiframe.

### 26) What do you understand by Mobile IP?

Mobile IP or MIP is an IETF (Internet Engineering Task Force) standard communications protocol designed to allow mobile devices' (such as laptops, PDA, mobile phones, etc.) users to move from one network to another while maintaining their permanent IP (Internet Protocol) address.

### 27) Why Mobile IP is so important?

Mobile IP is very important in Mobile communication because it facilitates mobile devices' users to move from one network to another without losing connectivity.

### 28) What are the different components of the Mobile IP?

The Mobile IP has the following main components:

**Mobile Node:** The mobile node is an end-user system or device with inbuilt software to provide network roaming capabilities. Examples of Mobile nodes are smartphones, cell phones, PDA (Personal Digital Assistant), laptop, Tablet etc.

**Home Agent:** The home agents are used to provide several services for the mobile node and are located in the home network. It maintains the location registry and acts as a starting point to send packets towards the mobile nodes.

**Foreign Agent:** The foreign agent acts as a default router for the mobile nodes. It provides several services to the mobile node during its visit to the foreign network.

**Care-of Address (COA):** It defines the mobile node's current location from an IP point of view. The packet delivery toward the mobile node is done using a tunnel, and the COA denotes the endpoint of the tunnel. In other words, you can say that it specifies the address where packets exit the tunnel.

**Correspondent Node (CN):** The correspondent node acts as a communication partner for the mobile node. It can be a fixed or mobile node.

**Home Network:** The home network is the subset of the mobile node. No mobile IP support is needed within this network.

**Foreign network:** A foreign network specifies a network in which a mobile node operates when away from its home network.

### 29) What is GSM in Mobile Computing?

GSM is an acronym that stands for Global System for Mobile Communications (GSM). It is a standard developed by the European Telecommunications Standards Institute (ETSI) to describe the protocols for second-generation (2G) digital cellular networks used by mobile devices.

### 30) In which year and country GSM was first deployed?

The GSM technology was first deployed in Finland in December 1991. By the mid-2010s, it became a global standard for mobile communications and covered over 90% market share. Now, it is operated in over 193 countries and territories.

### 31) Explain the GSM System Protocol Architecture.

The GSM System protocol architecture can be defined as a layered model that allows communications between two different systems. The lower layers provide the services to the upper-layer protocols. Each layer passes suitable notifications to ensure the transmitted data has been formatted, transmitted, and received accurately.

See the following diagram of the GSM System protocol architecture:

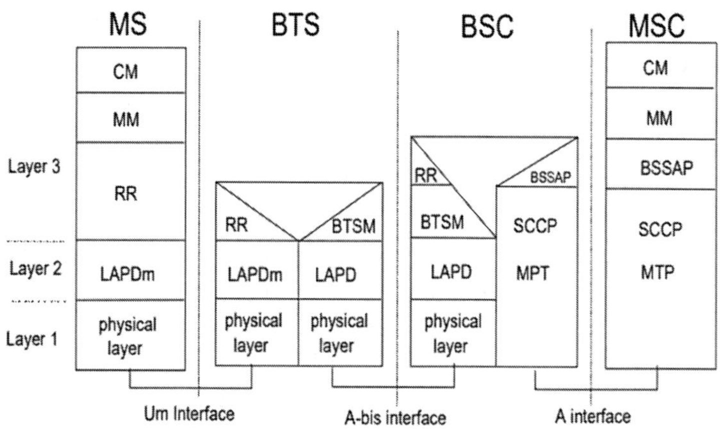


The GSM System protocol architecture consists of the following protocols:

**MS Protocols**

According to the architecture, the MS protocol is assembled into three general layers:

1. **Layer 1:** The layer 1 specifies the physical layer. It uses the channel structures over the air interface.

2. **Layer 2:** The layer 2 specifies the data-link layer. The data-link layer is a modified version of the Link access protocol for the D channel (LAP-D) protocol used in ISDN, called Link access protocol on the Dm channel (LAPDm).

3. **Layer 3:** The layer 3 specifies the GSM signaling protocol. it is divided into three sub-layers:

- Radio Resource Management (RR)
- Mobility Management (MM)
- Connection Management (CM)

### 32) What do you understand by TCP?

In Mobile computing, TCP stands for Transmission Control Protocol. It is a connection-oriented communication protocol that facilitates to exchange of messages between computing devices in a network. TCP is the most common protocol in networks. It uses the Internet Protocol (IP) to make the communication possible. That's why, together they are referred to as TCP/IP.

### 33) Explain the working of TCP in Mobile Computing.

The working of TCP can be defined in the following steps:

- TCP takes the messages from an application/server and divides them into packets.
- Now, the packets are forwarded by the devices in the network i.e., switches, routers, security gateways etc. to the destination.
- TCP assigns a number to each packet and reassembles them before handing them off to the application/server recipient.
- It is connection-oriented, so it ensures that a connection is established and maintained until the packet exchange between the application/servers sending and receiving is complete successfully.

### 34) What do you understand by OVSF?

OVSF is an acronym that stands for Orthogonal Variable Spreading Factor. It is used to separate the different data streams of a sender in UMTS by using orthogonal codes.

### 35) What is polling in Mobile Communication?

In Mobile communication, "polling" is a process of continuous checking of other programs or devices by one program or device to check their state. It usually sees whether they are still connected or want to communicate. In polling, there is one master station and several slave stations.

### 36) What do you understand by BSSGP in Mobile computing?

BSSGP stands for Base Station Subsystem GPRS Protocol. It is used to convey routing and QoS-related information between the BSS and SGSN. BSSGP does not perform error correction and works on top of a frame relay (FR) network.

### 37) What do you understand by WLAN?

WLAN is an acronym that stands for Wireless Local Area Network. It is a wireless distribution method for two or more devices that use high-frequency radio waves and often include an access point to the Internet. WLAN is the core concept of mobile communication. It facilitates users to move freely within the coverage area, such as a home or a small office while maintaining a network connection. Sometimes, WLAN is also called a local area wireless network (LAWN).

### 38) What are the advantages and disadvantages of WLAN?

Following are the advantages and disadvantages of WLAN:

#### Advantages of WLAN

**The wireless approach provides Flexibility:** The most significant advantage of a WLAN is that it provides a facility that devices can be connected wirelessly. It doesn't require cables, so it provides flexibility because even small devices such as smartphones and tablets can be connected to the network without any wire.

**Reduce Cost:** WLANs were very expensive in the early 1990s and were only used when wired connections were strategically impossible. But nowadays, the use of WLAN reduces the cost of the setup because it is cheap and allows homes and businesses to create local networks without wiring the building with Ethernet. The cost of cable and the wiring cost can be cut easily by using WLAN.

If you want to extend the connection area, you can easily upgrade WLAN by replacing routers with new versions of devices with higher capacity. It is a more comfortable and cheaper solution than upgrading old Ethernet cables.

**Support Unlimited Devices:** WLANs are not limited by the number of physical ports on the router so, you can connect dozens or even hundreds of devices. You can easily extend the range of a WLAN by adding one or more repeaters.

**Robustness:** It is robust because there is no chance of connection failure due to loose wire.

**Disadvantages of WLAN**

**Security:** Wireless networks (WLAN) are naturally less secure than wired networks. It is very important to limit access to the network if security is a concern because any wireless device can attempt to connect to a WLAN. You can use wireless authentication such as WEP or WPA, which encrypts communication.

**Susceptible to Interference:** WLAN is more susceptible to interference from other signals or physical barriers, such as concrete walls. That's why WLANs are not preferred over LANs. LANs offer the highest performance and security and are still used for many corporate and government networks.

**Quality of Service:** The quality of service in WLAN is more inferior than LAN.

**Low Speed:** WLAN doesn't provide as higher speed as LAN can provide.

**39) What is the difference between infrastructure and ad-hoc networks?**

The Difference between Infrastructure and Ad-hoc Networks:

| Infrastructure Network | Ad-hoc network |
|---|---|
| It is an infrastructure-based wireless network. | It does not require any infrastructure to work. |
| In this network, the communication takes place between the wireless nodes and the access point. | In this network model, each node can communicate directly with other nodes, so it doesn't require any access point controlling medium. |

**40) What are the main features of infrared transmission?**

Infrared transmission is used in the region of the electromagnetic radiation spectrum. Its wavelength is longer than the wavelength of visible light but shorter than the wavelength of radio waves.

The Infrared frequencies are higher than the frequencies of microwaves but lower than the frequencies of visible light.

### 41) Where is infrared transmission used?

The Infrared data transmission is generally used to connect laptops, smartphones, Personal Digital Assistants (PDAs) to peripheral devices such as printers or desktop computers for synchronizing work files.

### 42) In which applications the infrared waves are used in Mobile communication?

The infrared waves are used in the following applications of mobile communication:

- Remote controls for television, ACs, stereo and other home appliances
- Wireless peripheral devices such as modem, keyboard, mouse, printer etc
- Wireless LANs
- Fire detectors
- Night vision systems
- Intrusion detection systems
- Motion detectors and sensors

### 43) Describe the types of wireless applications.

Now a day mobile application is used in every field in real life. Some of them are.

- In Vehicles
- In Hospitals
- In Business
- In Entertainment
- In Location dependent services
- In Mobile and wireless devices
- Sensor
- Embedded controllers
- Pager
- Mobile phones
- Personal digital assistant
- Laptop

## 44) What are the limitation of mobile computing?

**Range and Bandwidth:** In general direct cable connection is faster than mobile internet conection. 2G, 3G, 4G networks are fast but these networks are usually available within range of cell phone towers. The main problem with wireless networks is that it works within limited range.

**Security standards:** When working mobile, user will use virtual private network (VPN). VPN should be carefully used. One can easily attack the VPN through a huge number of networks interconnected through the line.

**Power consumption:** if a power supply fails or portable generator is not available then mobile devices are relying on battery power. These batteries are more expensive.

**Transmission interferences:** The geographical area is not same everywhere. So the terrain can interfere with signal reception. The range from the nearest signal point may also interfere with signal reception. Signals in tunnels, some buildings, and rural areas are often poor.

**Health problem due to Cell Phone:** Mobile phone users may use phone while driving so it may cause accident. Cell phones may interfere with sensitive medical devices.

## 45) Describe cell in detail.

The geographic area or cellular service area is divided into small hexagonal region called cells. It is the basic unit of a cellular system. These cells collectively provide coverage over larger geographical areas. That is why a user can communicate through mobile phones even if the equipment is moving through cells during transmission. Each cell served by its own antenna. In cellular network adjacent cells are assigned different frequencies to avoid interference.

Advantages of cellular systems with small cells are the following:

- Higher Capacity
- Less transmission power
- Local interference only
- Robust

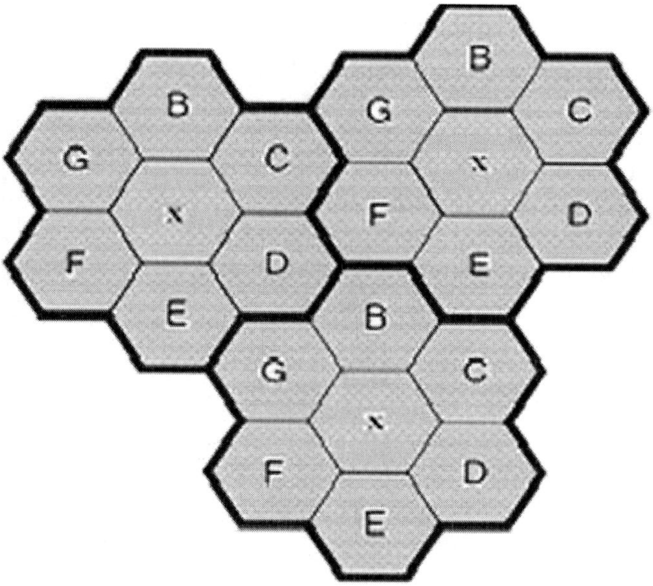


Fig: Hexagonal Cells

### 46) Describe Handoff in cellular networks.

When a user talks on the mobile phone to other user it may happen that the mobile station moves from one cell to another. During this conversation signal may become weak. To solve this problem, the Mobile Switching Center (MSC) checks the level of the signal every few seconds. If the strength of the signal is week then the MSC searches a new cell that can provide better communication.

Handoff is the process by which a mobile telephone call is transferred from one base station to another base station. There are two types of handoff.

- Hard Handoff
- Soft Handoff

Hard Handoff: In this process a mobile station only communicates with one base station. When the mobile user moves from one cell to another, communication must first be broken with the previous base station before communication can be established with the new one.

Soft Handoff: In this process a mobile station can communicate with two base stations at the same time. This means that a mobile station may continue communicate with the new base station before link is break off from the old one.

### 47) What is mobile agent?

A mobile agent is a software that can move across the network and represent various tasks. It is a distributed computing paradigm. So you can think agent is an independent software program that runs on behalf of a network user. When mobile agent program is launched by a user then it can travel from node to node autonomously, and can continue to function even if the user is disconnected from the network. Mobile agent works in different type of application and area such as: Internet, Electronic commerce, Mobile Computing, Networking.

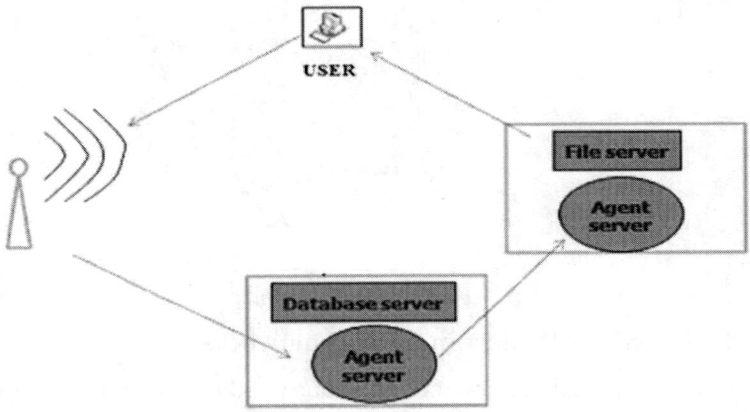


Fig: Working of Mobile Agent

### 48) What are the different services provided by GSM.

GSM provides the following types of services.

- Bearer services:
- Tele Service:
  1. Encrypted voice transmission
  2. Short message service
  3. Emergency number
  4. Data Communication
- Supplementary services:
  1. Useridentification
  2. call redirection
  3. Closed user groups

Bearer services permit transparent and non-transparent, synchronous or asynchronous data transmission.

### 49) Explain GSM architecture in detail.

A GSM system consists of three subsystems, the radio sub system (RSS), the network and switching subsystem (NSS), and the operation subsystem (OSS).

### 1. Radio subsystem

The radio subsystem (RSS) contains all radio specific entities such as the mobile stations (MS) and the base station subsystem (BSS). It also contains BTS and BSC.

Mobile Station

The MS comprises all user equipment (Cell Phone) and software required for communication with a GSM network. Cell Phone contains the subscriber identity module (SIM), which stores all user-specific data that is relevant to GSM network.

Base station subsystem (BSS): A GSM network contains many BSSs, each controlled by a base station controller (BSC).The main function of BSS are as follows:

- Establishing connection to MS
- Coding/decoding of voice

Base transceiver station (BTS): A BTS comprises of antennas, signal processing, and amplifiers necessary for radio transmission.

Base station controller (BSC): The BSC controls all the BTS. The main function of BSC are as follows:

- Reserves radio frequencies.
- Handles the handover.
- Performs paging of the MS.

### 2. Network and switching subsystem (NSS).

The NSS comprises MSC, HLR and VLR.

Mobile services switching center (MSC): These are high performance digital switches. An MSC manages several BSCs and provide connections to other MSCs and to the BSCs via the A interface. Home location register (HLR): It contains all the user related data such as the mobile subscriber ISDN number (MSISDN), subscribed services (e.g., call forwarding, roaming restrictions, GPRS), and the international mobile subscriber identity (IMSI).

Visitor location register (VLR): It is a dynamic database which stores all important data needed for the Mobile Station users currently in the (location area) LA that is associated to the Mobile Switching Center.

### 3. Operation subsystem.

It provides the necessary functions for network operation and maintenance. The OSS contains the following interties.

- Operation and maintenance center (OMC):
- Authentication Centre (AuC):
- Equipment identity register (EIR):

## 50) Describe Localization and calling in GSM

Worldwide localization of users and roaming are the main service provided by the GSM network system. The system always knows where a user currently is, and the same phone number is valid worldwide. For providing this service GSM updates the user location periodically. The HLR always contains information about the current location. VLR responsible for the MS informs the HLR about location changes. As soon as an MS moves into the new location area (range of new VLR), the HLR sends all user information needed to the new VLR.

To locate an MS and to address the MS, following numbers are needed:

- Mobile station international ISDN number (MSISDN).
- International mobile subscriber identity (IMSI).
- Temporary mobile subscriber identity (TMSI).
- Mobile station7 roaming number (MSRN).

## 51) Describe Mobile IP.

Mobile IP is an open standard that allows users to keep the same IP address and stay connected while roaming between IP networks. Mobile IP provides the facility to users to keep the same IP address while traveling to a different network. This ensures that roaming of users could continue communication without sessions or connections being terminated.

Components of a Mobile IP Network

Mobile IP has the following three components:

- Mobile Node
- Home Agent
- Foreign Agent

The Mobile Node is a device such as a cell phone or laptop. These devices have software that enables network roaming capabilities.

The Home Agent is a router on the home network that maintains information about the device's current location. It uses tunneling mechanism to forward packets from a device on the Internet called a Correspondent Node.

The Foreign Agent is a router that works as the connection point for the Mobile Node when it roams to a foreign network, delivering packets from the Home Agent to the Mobile Node.

### 52) What is Hidden terminals problem?

Suppose that there are three mobile A, B, and C. The transmission range of A reaches B, but not C. The transmission range of C reaches B, but not A. Finally the transmission range of B reaches A and C.

A starts sending signal to B, C does not receive this transmission. C also wants to send data to B and senses the medium. The medium appears to be free, the carrier sense fails. C also starts sending data. It will create a collision at B. But A cannot detect this collision at B and continues with its transmission. A is hidden for C and vice versa.

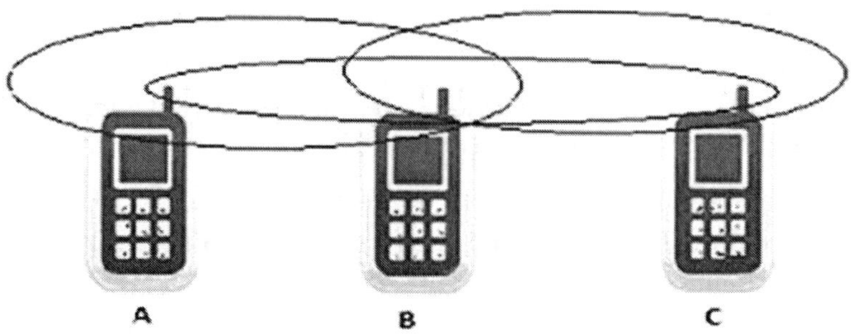


### 53) Give the difference between the network 1G,2G,2.5G,3G mobile communication?

1G - Voice-only communication. 2G – Communicate voice as well as data signals. 2.5G – Enhancements of the second generation and sport data rates up to 100 kpbs. 3G – Mobile devices communicate at even higher data rates and support voice, data , and multimedia streams. High data rates in 3G devices enable transfer of video clips and faster multimedia communication.

### 54) What are the multiplexing techniques ?

The Multiplexing techniques are: i)Space division multiplexing. ii) Time division multiplexing. iii)Frequency division multiplexing. iv)Code division multiplexing.

### 55 ) Define Space Division Multiplexing Access?

Space division multiple access (SDMA) means division of the available space so that multiple sources can access the medium at the same time. SDMA is the technique in which a wireless transmitter transmits the modulated signals and accesses a space slot and another transmitter accesses another space slot such that signals from both can propagate in two separate spaces in the medium without affecting each other.

### 56 ) Define Code division multiplexing Access?

CDMA(Code Division Multiple Access) is an access method in which multiple users are allotted different codes (sequence of symbols) to access the same channel (set of frequencies)

### 57) Define Time division multiplexing Access?

Time division multiplexing (TDMA) is an access method in which multiple users, data services, or sources are allotted different time-slices to access the same channel. The available time-slice is divided among multiple modulated-signal sources. These sources use the same medium, the same set of frequencies, and the same channel for transmission of data.

### 58) Define Frequency division multiplexing Access?

Frequency division multiple access (FDMA) is an access method in which entails assignments of different frequency-slices to different users for accessing the same carrier.

### 59) What is the aim of ubiquitous computing?

- The aim of ubiquitous computing is to design computing infrastructures in such a manner that they integrate seamlessly with the environment and become almost invisible.

- Present Everywhere Bringing mobile, wireless and sensor Ubiquitous computing (ubicomp) integrates computation into the environment, rather than having computers which are distinct objects

### 60) What are the characteristics and key constraints of mobile computing devices?

Characteristics of mobile computing

- Adaptation Data dissemination and Management
- Heterogeneity Interoperability Context awareness

Key constraints of mobile computing

- unpredictable variation in network quality
- lowered trust and robustness of mobile elements